# FLYING SCOTSMAN

First published in a larger format as *Flying Scotsman Owners' Workshop
Manual* in 2016 and reprinted in 2016, 2017, 2018 and 2019
This edition published in 2020

A catalogue record for this book is available from the British Library

ISBN 978 1 78521 689 3

Library of Congress control no. 2019948380

Published by Haynes Publishing,
Sparkford, Yeovil,
Somerset BA22 7JJ, UK.
Tel: 01963 440635
Int. tel: +44 1963 440635
Website: www.haynes.com

Haynes North America Inc.,
859 Lawrence Drive, Newbury Park,
California 91320, USA.

Printed in Malaysia.

National Railway Museum logo © SCMG Enterprises Ltd.

This book is produced in association with the
National Railway Museum, York.
Sales of this book help support the museum's exhibitions
and programmes.
www.nrm.org.uk

All images © National Railway Museum/
Science & Society Picture Library, unless otherwise stated.

# FLYING SCOTSMAN

**Philip Atkins**

HAYNES **ICONS**

**RAIL**WAY
**MUS**EUM

# Contents

# Introduction

One of the first books that I was ever aware of was my late father's copy of O. S. Nock's *The Locomotives of Sir Nigel Gresley*, which was published immediately after the war. During the 1930s Dad had extensively photographed Gresley A1 and A3 (but not A4) 4-6-2s in their prime, including the legendary LNER No. 4472 *Flying Scotsman*, complete with 'Flying Scotsman' headboard, as it paused at Grantham in 1933 (see right).

As a small boy in the early 1950s, my own early railway memories include family picnics enjoyed in a field at South Muskham, just north of Newark, beside the water troughs on the East Coast Main Line (ECML). At that time there were no fewer than 202 named three cylinder 4-6-2s of Doncaster origin in service, which were officially comprised of eight different classes, most of which worked on the ECML. My own abiding memory is of streamlined A4s bearing down from the north, with their overfilling tenders throwing out spectacular spray as they sped past. I might even have also glimpsed the by then thirty-year-old *Flying Scotsman*, at that time British Railways No. 60103, but my first certain memory of encountering it was in April 1960, when I saw it turning on the short-lived track triangle at Grantham shed. Oddly enough, Dad and I unexpectedly saw it again much nearer home in Nottingham only two days later, passing through New Basford on the former Great Central Main Line, deputising for the preserved GWR 4-4-0 *City of Truro* on a society excursion. Just three years later, in May 1963, we both witnessed it once more, a mile or two further north at Bulwell Common, on one of its first outings after its very recent restoration to pre-war LNER apple green glory.

I am delighted that the long awaited restoration of *Flying Scotsman* has now been completed, and that it will be travelling extensively throughout the country.

**Philip Atkins, Harrogate, February 2016**

## Acknowledgements

This book has only been possible through the assistance of many people, which includes several friends and former colleagues at the National Railway Museum, York, i.e. Peter Alston,

Karen Baker, Ed Bartholomew, Chris Beet, Charlie Bird, Dave Burrows, Chris Chesney, Tony Filby, Lorna Frost, Bob Gwynne, Chris Hogg, Lynn Patrick, Dave Sample, and Andrew Scott. Also Andy Croxton, Peter Thorpe, Bernie Walton, Mary White and Dawn Whitehead of the National Railway Museum Search Engine staff.

Lynn Patrick was responsible for much of the very fine photographic record of No. 4472, both when operating from York during 2004–05 under early National Railway Museum ownership, and while subsequently undergoing repairs at York and Bury, and to the boiler in Devon. Following Lynn's departure from the National Railway Museum in late 2011, the photographic record has been continued by Allan Baker, Mike Corbett, Peter Heaton and Paul Thompson. The writer would also like to express his indebtedness to Bill Andrew, Allan Baker, Richard Carmen, Chris Nettleton (of the Gresley Society), Dave Rollin, Alan Wappat, and to Chris Birmingham and Colin Green at Bury, both for invaluable technical advice and photographic assistance. Finally, to my wife Christine who rendered considerable technical assistance in the preparation of this book.

## Author's note

In the interests of brevity, and to avoid ambiguity, the subject of this book will usually be referred to hereafter simply as No. 4472, as the name 'Flying Scotsman' was applied to both the locomotive itself, and to the famous railway passenger service between London and Edinburgh, which the former worked *regularly* for a period of only eight years.

**ABOVE** Halcyon days. LNER 4-6-2 No. 4472 *Flying Scotsman* in its prime on the up 'Flying Scotsman' pausing at Grantham in 1933, a photograph taken by the author's father. *(G.H.F. Atkins)*

# CHAPTER ONE

# Setting the scene

The dawn of a legend. Newly named *Flying Scotsman* and renumbered 4472, Britain's sixth 4-6-2 locomotive is resplendent in the special finish applied in early 1924 for the British Empire Exhibition at Wembley.

# Identity crisis?

The London & North Eastern Railway 4-6-2 steam locomotive No. 4472 *Flying Scotsman,* is undoubtedly, individually, the most famous of the *c*110,000 steam locomotives to have been built in Great Britain. It is also arguably, throughout the world too, wherein an estimated 650,000 steam locomotives have been built over a period of more than 200 years, between 1804 and 2014.

Built by the Great Northern Railway, this locomotive entered service and ran during 1923 carrying its GNR number of 1472, although lettered L&NER. Somewhat prosaically, it owed its subsequent legendary status to having very soon sustained a broken middle piston rod. This could not immediately be replaced, which directly led to its selection for display at the British Empire Exhibition at Wembley in 1924, for which it was bestowed with its memorable appellation. Before this event the engine was truly anonymous. The *North Eastern Magazine* for March 1924 reproduced a photograph of it suspended beneath a crane, which was simply captioned: 'LNE "Pacific" for Wembley Exhibition being lowered on to wheels in Doncaster Shops'.

No. 4472's glory years were undoubtedly 1928–36, merely *one fifth* of its ordinary working life, and the only period when *Flying Scotsman* would have been seen in Scotland! During this time it was equipped with a special corridor tender to operate the LNER's flagship non-stop (summertime only) namesake 'Flying Scotsman'

express passenger service between the English and Scottish capital cities. From October 1936 it ran with a conventional tender again until withdrawal just three weeks short of its 40th anniversary, having accrued an estimated 2 million miles to its credit, which was some 600,000 miles behind that achieved by its close contemporary, No. 4475 *Flying Fox.*

By this time, very little of the original 1923-built engine still remained. Over the course of 40 years the engine had carried 15 different boilers, the last fitted only a few months before withdrawal, and it had operated with four different tenders of three distinct designs. In April 1928 alone, after only five low-mileage years, No. 4472 had received a new boiler, replacement outside cylinders, and a brand-new tender. The main frames would later have been renewed, probably after only ten years, and the cylinders were repeatedly replaced, so what did constitute its identity?

The Doncaster Works philosophy was that the identity of a locomotive was the number (which happened to be) on the cabside. But even this could change, and in the immediate post-war years it frequently did. In fact, No. 4472 successively carried *six* different running numbers in all, i.e. 1472, 4472, 502, 103, E103, and finally, 60103. These were for periods ranging from four months to 22 years. It was successively painted in apple green, black, apple green again, blue and finally, dark (Great Western!) green, having been spared the early British Railways 'purple' briefly applied to a few other A3s in 1948.

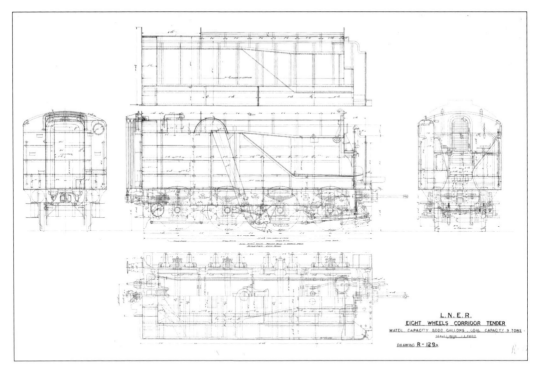

L. N. E. R.
EIGHT WHEELS CORRIDOR TENDER
WATER CAPACITY 5000 GALLONS · COAL CAPACITY 9 TONS ·
SCALE 1IN/1FOOT
DRAWING R-129a

Just four years before retirement its appearance was transformed by the fitting of a double chimney, which in turn prompted the provision of distinctive German-style 'trough' smoke deflector plates two years later.

On retirement, No. 4472 was purchased privately by the late Alan Pegler, who had No. 4472 restored to its pre-war apple green livery and reunited with a corridor tender. However, while so matched during 1928–36, it had not operated as an A3 with the later-pattern higher pressure boiler, with its distinguishing square 'blisters' on each side of the smokebox. These were only acquired during the final months of LNER ownership in 1947, following conversion from Class A1 to A3, and after it had been renumbered to 103.

Also, No. 4472 had not sported the highly distinctive 'banjo'-shaped dome casing of the later A3 boilers, even when it was lettered 'LNER'. It only gained this particular feature in early 1948, at the dawn of a new era, when its tender was lettered 'British Railways' and it was temporarily renumbered E103. Finally, as BR No. 60103, the engine was belatedly converted from right- to left-hand drive, as late as 1954.

Such anomalies were fully recognised by Alan Pegler. He later philosophically observed: 'It was really out of the question to consider trying to convert [No. 4472] to her original [A1] condition and I was prepared to settle for a "typical A3" as a reasonable compromise.' In fact, only two A3s with 'banjo' domes (*Salmon Trout* and *Brown Jack*), ever ran paired with corridor tenders, and then only for a couple of years.

Forty years later No. 4472 was in an even more anomalous configuration. In 1978, it had been fitted with an A4-type boiler, a pattern which the engine had not previously carried, although this was not immediately obvious, while in 1993, it reverted to its ultimate 1959 double-chimney plus 1961 smoke deflectors-mode. Far more appropriate to this new guise, however, it was also repainted in BR dark green and numbered 60103. Yet in 1999, while still retaining its '1962' physical configuration, it was nevertheless returned to LNER apple green as No. 4472 once again. It was in this confused format that the locomotive was purchased by the National Railway Museum five years later.

**ABOVE** Sectional general arrangement drawing (R-129) for the original 1928 series of corridor tenders, specially built for the new London–Edinburgh non-stop workings. This type of tender was only fitted to No. 4472 in LNER time between 1928 and 1936, but appeared again, in slightly modified form, throughout its 'second life' from 1963 onwards.

# Pacifics

The 4-6-2 tender engine, or 'Pacific', constituted the undoubted elite of British steam locomotive power. A total of 466 4-6-2s were built for service in Great Britain between 1908 and 1954, of which 458 were simultaneously in service on British Railways throughout 1955-58. No. 4472 was only the sixth to enter service; a year earlier there had been only one, the GWR *Great Bear*.

Remarkably, all but 20 of these 4-6-2s were constructed by the railways themselves. However, several hundred Pacifics were built for export between 1903 (3ft 6in gauge for South Africa) and 1955 (5ft 6in gauge for India), by the private British locomotive industry, which was dominated by the North British Locomotive Company in Glasgow (NBL). In 1924, NBL built 20 near repeats of No. 4472 as originally turned out, according to a detailed printed specification drawn up by the LNER in November 1923 from which extracts are reproduced in Chapter 5. The order was placed with NBL the following month.

## Quality machines

In recent years, in particular, the steam locomotive has sometimes been portrayed as a crude machine, but in fact, it was one of necessarily considerable precision, built to tolerances of often just a few thousandths of an inch. It was finely crafted with considerable skill from cast iron, steel, copper and brass, whose very qualities in themselves were precisely defined and their approved manufacturers specifically identified. (See Appendix 7.)

## Works photographs and drawings

No. 4472 (as would be GNR No. 1472) was not photographed during its construction in late 1922, nor even when newly completed as the first 4-6-2 to be lettered L(&)NER. Doncaster Works had a tradition of making a serial photographic record of the construction process of a new locomotive type, and so had chronicled the genesis of GNR No. 1470 just a year earlier. These images, from the National

Railway Museum Photographic Archive, effectively show just how No. 1472, the third Doncaster 4-6-2, would have appeared while building and clearly illustrate several of its major components 'in the raw'. These have been supplemented by numerous digital photographs taken by National Railway Museum staff to record No. 4472's 2004/5 operations and subsequent very extensive repairs. The National Railway Museum also holds many of the original working drawings for each of the various Doncaster 4-6-2 designs, from the Gresley A1 of 1922 to the Peppercorn A1 of 1948.

## Who drove *Flying Scotsman*?

Over many years, the National Railway Museum at York has received numerous enquiries from members of the public seeking confirmation, or actual documentary evidence, that one of their ancestors once drove LNER No. 4472 *Flying Scotsman*. The disappointing but simple answer is that no official records were kept linking any particular engineman to a specific locomotive, even though, pre-1939, in the case of 'top link' locomotives, these were often in practice handled by regular crews.

Many different drivers would in fact have held the regulator on No. 4472 over the years, of whom probably the best known were Albert Pibworth and Bill Sparshatt.

Drivers in the 'Top Link' responsible for the non-stop 'Flying Scotsman' workings in 1928, who would therefore have driven No. 4472 in its prime, were identified in the *Railway Magazine* for June 1928 as being:

- ■ **King's Cross**: A. Pibworth, B. Glasgow, J. Day and H. Miles.
- ■ **Gateshead**: T. Blades, H. Pennington, J. Gascoigne, J. W. Halford, J. G. Smith and J. G. Eltringham.
- ■ **Edinburgh Haymarket**: T. Henderson, T. Roper, R. Sheddon and T. Smith.

Railway personnel records for England and Wales are held by the National Archives at Kew, which are filed by personality on a seniority basis and not by location, for example, by locomotive depot in the case of locomotive crew.

# CHAPTER TWO

# 'Flying Scotsman' – the train

No. 4472 ascends Holloway bank, north of King's Cross, with the down non-stop 'Flying Scotsman', c1930. A brake third vehicle is marshalled with the passenger compartments next to the tender, to accommodate the relief engine crew.

## Up to 1939

The term 'Flying Scotsman' came to particular prominence in 1924 when No. 4472 was so named for display at the British Empire Exhibition at Wembley. The title had been used unofficially for around 50 years as the name of the long-established London–Edinburgh, Edinburgh–London expresses, which began running in 1862 under the name of the 'Special Scotch Express', but in 1924 the LNER officially renamed the service the 'Flying Scotsman'. The expresses traditionally departed simultaneously from each capital city at 10am (or thereabouts).

When Gresley 4-6-2s first took over regular working of the 'Scotsman' between London and York in the summer of 1924, an 8½hr schedule was still in force, which corresponded to an average speed of only 46mph. This did not necessarily preclude high speeds en route, however. The *Railway Gazette* for 14 November 1924 recorded that an unidentified 4-6-2 on a 12-coach down train, aggregating 356 tons, had recently attained 89mph on the gently falling gradient between Biggleswade and Sandy.

This was probably about the maximum speed possible with the original short-travel piston valves. The pedestrian 8½hr timing dated back 30 years to the aftermath of the so-called London–Aberdeen 'railway races' of August 1895. These had actually shown that a 6½hr schedule (London–Edinburgh, albeit with much lighter trains), inclusive of stops of only three to five minutes each, for locomotive changing purposes at Grantham, York, and Newcastle (and corresponding to a remarkably high overall average speed of 60.5mph), was technically possible even then.

However, public disquiet as to possible recklessness on the part of the railway companies resulted in a mutual agreement between the respective East Coast and West Coast railway partners whereby they agreed to adhere to a leisurely 8½hr schedule. In May 1932 this long-outmoded impediment was finally breached by the LNER to bring the time down to 7½hr, raising the overall average speed from a pedestrian 46.3mph to a still fairly modest 52.4mph.

## After 1945

Non-stop operation (after 1928 throughout July–August only) of the 'Flying Scotsman' was restored by the newly established British Railways in June 1948, only for it to be almost immediately temporarily diverted from Newcastle to Carlisle, and thence over the Waverley Route as far as St Boswell's and thence across to Tweedmouth. This was caused by exceptional flood damage to the ECML near Grantshouse during the following August and it briefly extended the non-stop working distance by nearly 16 miles to 408½ miles. However, this unique feature of the summer 'Flying Scotsman' was quickly transferred in 1949 to the new 'Capitals Limited' service with its 9.30am departures. This was later re-named 'The Elizabethan' in recognition of the Coronation in 1953. Thereafter, both the up and down 'Scotsman' made a stop at Newcastle for locomotive changing. Despite the arrival of the new Peppercorn A1s during 1948–49, the older A3s sometimes still worked the illustrious train in the 1950s, and even occasionally in the early 1960s, when diesel-electric traction was rapidly taking over.

During its final months in late 1947, the LNER obtained quotations for the supply of 25 1,600hp diesel-electric locomotives, to be used in pairs on its principal Anglo-Scottish passenger services, including the 'Flying Scotsman'. However, following the formation of British Railways on 1 January 1948, this ambitious scheme was quietly and quickly forgotten, thereby postponing dieselisation of the East Coast Main Line for several years.

The now legendary 3,300hp English Electric 'Deltics', introduced one year later than originally planned in 1961 (and initially only intended as a stop-gap pending the then anticipated 'imminent' electrification of the East Coast Main Line by 1970), were designed for 100mph running from the outset. In the event these were superseded in 1978 by the highly successful InterCity 125 High Speed Trains (capable of in excess of 140mph). With the inauguration of long-overdue electrification 12 years later in 1990, the London–Edinburgh time was reduced to just over four hours, to give a once-undreamed of average speed in the order of 95mph. In May 2011, the 'Flying Scotsman' service was re-launched by East Coast in the up direction only, when the journey time was further reduced to 4 hours, departing Edinburgh at 05.40 with a single stop at Newcastle.

# Rolling stock

Prior to 1923, the rolling stock employed on through trains between London and Edinburgh was proportionately, according to their relative contributed mileage along the East Coast Main Line (or ECML), jointly owned by the Great Northern, North Eastern and North British railway companies. Attractively finished in varnished teak, the coaches were lettered in gold either 'ECJS' or 'East Coast Joint Stock' in full. Initially, the stock was only box-like four-wheelers, and during the 1880s it comprised exclusively rigid-frame six-wheelers. From 1893, bogie vehicles, both eight-wheel and 12-wheel, increasingly dominated. These vehicles were principally built at Doncaster (by the GNR) and York (by the NER), with a small number by the NBR at Cowlairs (Glasgow).

The LNER built new varnished teak-bodied stock for the 'Flying Scotsman' service in 1924. This incorporated existing full brakes at each end, but their formation was amended with the advent of the non-stop service on 1 May 1928. The formation typically, now became: corridor 3rd brake, corridor 3rd, corridor composite (locker), corridor 3rd, buffet lounge, corridor 3rd, restaurant 3rd, kitchen car, restaurant 1st, corridor 1st, corridor 3rd, and luggage van. Variously, gentlemen's barbers, ladies' hairdressing facilities, and even a cocktail bar were provided on board. The landmarks along the route were detailed in the *On Either Side* guide published by the LNER, which helpfully identified the numerous cathedrals, castles and sites of former battles passed at speed.

By 1938, when further new teak stock was introduced, streamlined A4s had recently taken over the 'Flying Scotsman' workings. The non-stops were suspended during the Second World War, after which new Thompson steel-bodied stock began to be provided.

After more than 30 years, Pacific domination of the King's Cross–Edinburgh passenger services rapidly faded after 1960. However, a few months into the new six-hour diesel timing, the 'Scotsman' was again steam hauled, at least in part, appropriately by a

| Year | Motive power | Journey time (down train) |
|------|-------------|---------------------------|
| 1880 | GNR 8ft 4-2-2/NER '901' 2-4-0 | 9hr |
| 1910 | GNR 'large' 4-4-2/NER Class V 4-4-2 | 8½hr |
| **1925** | **LNER Class A1 4-6-2** | **8½hr** |
| **1932** | **LNER Class A1/A3 4-6-2** | **7½hr** |
| 1937 | LNER Class A4 4-6-2 | 7hr |
| 1946 | LNER Class A4 4-6-2 | 8hr |
| 1955 | LNER-type Class A1 (Peppercorn) 4-6-2/ A4 4-6-2 | 7hr |
| 1962 | BR English Electric 3,300hp Co-Co DE ('Deltic') | 6hr |
| 1977 | BR English Electric 3,300hp Co-Co DE ('Deltic') *After track improvements* | 5hr 27min |
| 1982 | BR High Speed Train 4,500hp DE (HST 125) | 4hr 35min |
| 1994 | BR Class 91 6,300hp Bo-Bo 25kV AC electric | 4hr 10min |

**JOURNEY TIMES AND TYPICAL MOTIVE POWER EMPLOYED ON THE 'FLYING SCOTSMAN' LONDON–EDINBURGH PASSENGER SERVICE, 1880–1994**

DE = diesel-electric

**BELOW** The prototype 6,300hp Class 91 Bo-Bo electric locomotive No. 91101 (originally 91001), built at Crewe Works in 1988, branded as *FLYING SCOTSMAN* (in LNER-style Gil Sans lettering) in May 2011 by East Coast Railways, seen at York on 6 July 2013. In October 2015 this locomotive was repainted in Virgin Trains East Coast livery and fitted with stainless steel nameplates embodying the Scottish thistle. When brand new, a sister Class 91 locomotive, No. 91010, attained 162mph while descending Stoke Bank. These locomotives have typically covered 1,000 miles per day, and will have accrued exceptional life mileages by the time they are due to be replaced from 2018. *(Author)*

Gresley A3. On 1 November 1962, 1923-built No. 60110 *Robert the Devil* was called upon at King's Cross to deputise for a failed 'Deltic' on the down train as far as Newcastle, which (excusably) was reached 35 minutes late. Just over a year later, on 9 December 1963, No. 60106 *Flying Fox* (also 1923-built) was reported to have worked the final leg of the up 'Scotsman' from Peterborough to King's Cross. So ended an era.

## The route of the 'Flying Scotsman' train

Although alternative direct routes between London and Edinburgh were investigated by Parliamentary committees as early as 1839, in practice, what eventually became known as the East Coast Main Line evolved in a piecemeal fashion over a period of many years. The eventual route, as finalised in 1906, totalled 393 miles, and was regularly traversed throughout by LNER 4-6-2 locomotives over a period of almost exactly 40 years (1923–63). Its evolution can be summarised as follows:

The route included four particularly notable stations: London King's Cross (opened 1852), York (1877), Newcastle Central (1850), and Edinburgh Waverley (in essentially its present form, 1900). It also passed over three major bridges/viaducts, i.e. Welwyn Viaduct across the Mimram Valley in Hertfordshire, the King Edward Bridge spanning the River Tyne between Gateshead and Newcastle, and the Royal Border Bridge crossing the River Tweed between Tweedmouth and Berwick-upon-Tweed.

Compared with its West Coast counterpart, running between London Euston and Glasgow, the East Coast Main Line was relatively undemanding, with gradients rarely exceeding 1-in-200. By far the most sustained is the five-mile 1-in-96 Cockburnspath bank (southbound) ascending to the Scottish summit at Grantshouse, north of Berwick. The highest point in England is Stoke Summit south of Grantham, only some 20ft less at 345ft above sea level, in the notably flat county of Lincolnshire.

By contrast, the corresponding twin summits on the West Coast Main Line at Beattock (1,033ft), and at Shap (915ft), both involve northbound

**BELOW RIGHT Simple map showing the East Coast Main Line.**

| Opened | |
|--------|--|
| 1852 | London, King's Cross–Maiden Lane (temporary London terminus) |
| 1850 | Maiden Lane–Werrington Junction (near Peterborough) |
| 1852 | Werrington Junction–Retford |
| 1849 | Retford–Doncaster |
| 1848 | Doncaster–Shaftholme Junction (limit of Great Northern Railway) |
| 1871 | Shaftholme Junction–Chaloners Whin Junction (near York) |
| 1839 | Chaloners Whin Junction–York |
| 1841 | York–Polam Junction (Darlington) |
| 1829 | Polam Junction–Parkgate Junction (Darlington) |
| 1844 | Parkgate Junction–Coxhoe Junction |
| 1871 | Coxhoe Junction–Relly Mill Junction (near Durham) |
| 1856 | Relly Mill Junction–Newton Hall Junction |
| 1868 | Newton Hall Junction–Gateshead West Junction |
| 1906 | Gateshead West Junction–Newcastle Central station |
| 1848 | Newcastle Central station–Manors |
| 1839 | Manors–Riverside Junction |
| 1847 | Riverside Junction–Tweedmouth |
| 1848 | Tweedmouth–Berwick-upon-Tweed (limit of North Eastern Railway) |
| 1846 | Berwick–Edinburgh Waverley (North British Railway) |

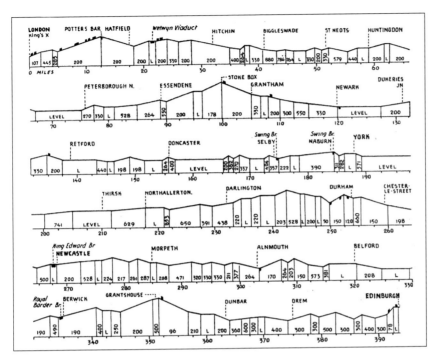

climbs of several miles at 1-in-75, which were also often associated with adverse weather conditions.

The East Coast Main Line has undergone significant changes since the end of the steam era, including improved track layouts at Peterborough, where a permanent speed restriction of only 20mph had long been in force. Following a serious structural collapse inside the Penmanshiel Tunnel, an impromptu bypass was hastily put in place during 1979, and the Selby diversion was completed in 1983. Colour light signalling, initiated pre-war, gradually superseded mechanical semaphores.

Full electrification of the East Coast Main Line, proposed in the British Railways Modernisation and Re-Equipment Plan announced in January 1955, was not authorised until 1984 and was completed in 1990. In 1919, the North Eastern Railway had proposed to electrify York–Newcastle (and possibly beyond, through to Edinburgh), even building a prototype 4-6-4 electric locomotive in 1922, although the corresponding infrastructure did not materialise.

Prior to the Grouping of 1923 the Great Northern Railway had traditionally changed locomotives at the old market town of Grantham (105.5 miles north of London), which then ran through to York (a further 82.7 miles), where North Eastern Railway locomotives took over. These in turn were changed at Newcastle (80.1 miles) for the longest single stretch (124.4 miles) on to Edinburgh, which included 57.5 miles over North British Railway metals north of Berwick. In 1928, a select band of 4-6-2s would begin to run almost 400 miles between London and Edinburgh without any scheduled stops en route, a unique operation which was finally terminated in September 1961.

**BELOW** The 1930s LNER England/ Scotland boundary marker at Marshall Meadows, 2½ miles north of Berwick-upon-Tweed, as amended by British Railways.

# CHAPTER THREE

# From Stirling 4-2-2 to Gresley 4-6-2

Formal 'shop grey' official portrait of the pioneer Gresley 4-6-2, GNR No. 1470 *Great Northern*, as completed at Doncaster Works in early 1922.

# Locomotive development

Three locomotive engineers presided over the locomotive department of the Great Northern Railway at Doncaster Works over a period of 56 years. These were Patrick Stirling, between 1866 and 1895, Henry Ivatt, during 1896–1911, and finally, Nigel Gresley, 1911–22. Under each of these very able personalities, locomotive development progressed with a continuity rarely found in other contemporary railway companies.

Performance-wise, Patrick Stirling's inside-cylinder 2-2-2s were allegedly superior, however, he has become immortalised by his highly distinctive outside-cylinder '8ft single' 4-2-2s, of which the pioneer, No. 1 of 1870, was set aside for preservation by the GNR as early as 1907. This was originally designed to handle 150-ton trains at average speeds of 50mph. It remained in intermittent production for 25 years, up to Stirling's death in office in 1895. During this time, the cylinders (which had the unusually long piston stroke for the period of 28in) were increased in bore from 18in to 19½in, boiler pressure was raised from 130lb to 160lb per sq in, and the driving axle load increased from only 15 to 19¼ tons, or covertly probably as high as 20 tons.

Even so, by 1895, the design was scarcely equal to meet rapidly increasing demands, with the recent advent of heavier bogie corridor passenger stock, sleeping and dining cars, all to be worked at progressively higher speeds. Therefore, double heading with two locomotives on a train was becoming a necessity. Between 1890 and 1896 alone, train weights had typically increased from 170 to 237 tons, and average speed, between London and York, from 51¼mph to 55¾mph. However, the actual passenger payload remained virtually unchanged at about 250.

Although Stirling's successor, Henry Ivatt, built the very last British 'single' of all (also a 4-2-2 but with inside cylinders) at Doncaster as late as 1901, on arrival he had quickly instituted the design of Britain's first 4-4-2, or 'Atlantic' to use American parlance. This reversed the small-boiler/large-cylinder characteristics of Stirling's 4-2-2s, for a large-boiler/small-cylinder policy. The prototype, No. 990, later named *Henry Oakley*, appeared in May 1898, and was thoroughly evaluated before a further 20 similar engines were turned out during 1900–02.

At the close of 1902, an additional 4-4-2, No. 251, was experimentally completed with a much larger, 5ft 6in diameter boiler with a wide firebox. Ninety 'large-boilered' 4-4-2s were then built at Doncaster up to 1910, the last examples appearing new with superheaters and piston valve cylinders, in line with very recent developments. GNR Nos 990 and 251 are now both preserved in the National Collection, as the only surviving examples of 300 4-4-2s built for seven British railway companies between 1898 and 1921.

# NIGEL GRESLEY

Herbert Nigel Gresley was born in Edinburgh on 19 June 1876, the fifth and youngest child of a Derbyshire clergyman and baronet. After education at Marlborough School in 1893 he entered Crewe Works as a premium apprentice of F.W. Webb on the London & North Western Railway, before moving to the Horwich locomotive drawing office of the Lancashire & Yorkshire Railway under J.A.F. Aspinall in 1898.

He then occupied a succession of positions on the LYR, which included moving to the rolling stock side and becoming assistant works manager at Newton Heath Works in 1900. With this valuable experience he transferred to the Great Northern Railway as carriage and wagon superintendent at Doncaster Works in 1905, under H.A. Ivatt, who was a close personal friend of Aspinall, who had doubtless recommended him. Gresley thereupon promptly abandoned the clerestory roof characteristic of Great Northern and East Coast Joint passenger stock in favour of the more modern elliptical construction.

In 1911, Gresley succeeded Ivatt to take overall charge of locomotive, carriage and wagon matters. His first entirely new locomotive design was a superheated mixed traffic 2-6-0, No. 1630, with outside cylinders and external Walschaerts valve gear, then something of an innovation, which appeared in August 1912. In 1913, he introduced a powerful 2-8-0 mineral engine with similar characteristics, of which a three-cylinder development was experimentally built later, in 1918. Particularly notable was a three-cylinder large-boilered 2-6-0 introduced in 1920, and large express passenger 4-6-2, in 1922. Except for two classes of inside-cylinder 0-6-0 goods engine, three cylinders became a trademark of all of Gresley's subsequent conventional locomotive designs.

Nigel Gresley's appointment as Chief Mechanical Engineer of the new London & North Eastern Railway was by no means a foregone conclusion on vesting day, 1 January 1923. John Robinson, late of the Great Central Railway, then 66 and therefore 20 years senior to Gresley, declined the offer of the post and personally recommended the latter instead, which Gresley assumed on 24 February. This was in fact, the same day upon which No. 4472 entered traffic! In 1925 Gresley experimentally introduced two 2-8-2 heavy mineral engines, directly derived from his recent express passenger 4-6-2, which in the event were not repeated.

In 1934, the outstanding French-influenced express passenger 2-8-2, No. 2001 *Cock o' the North*, with rotary cam poppet valve gear, was built specifically for service between Edinburgh and Aberdeen. This was followed by the streamlined A4 class 4-6-2 in 1935, and the V2 mixed traffic 2-6-2 in 1936. Gresley's last new locomotive design was the

ABOVE **Portrait of Nigel Gresley, c1915.**
*(Courtesy, The Gresley Society)*

lightweight V4 class 2-6-2, which made its wartime debut in early February 1941, only two months before his death.

Gresley had also made considerable innovations regarding rolling stock, including articulation, but unlike his contemporaries, continued to favour wooden (teak) carriage bodies in preference to steel panelling, well into the 1930s. He was awarded a knighthood in 1936, and his 100th 4-6-2, A4 No. 4498 (now preserved), when brand-new was named after him in December 1937. Large in physical stature, a notable attribute of the man was that he also 'thought big' and was highly receptive to bold new ideas, including several from overseas.

Between 1908 and 1930 he was granted 16 innovative patents relating to steam locomotives and railway rolling stock. A curious weakness, however, was his distinct reluctance to address any shortcomings which might become apparent after a new locomotive design had been introduced, such as was well exemplified by the 4-6-2 valve travel question, which is discussed in Chapter 4. Gresley died in harness at his home in Hertfordshire on 5 April 1941, two months short of his 65th birthday, when he had been due to retire. He is buried at Netherseal in south Derbyshire.

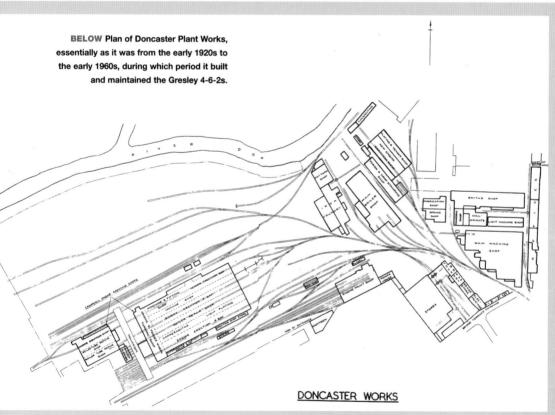

**BELOW** Plan of Doncaster Plant Works, essentially as it was from the early 1920s to the early 1960s, during which period it built and maintained the Gresley 4-6-2s.

DONCASTER WORKS

Ivatt retired in 1911, to be succeeded by Nigel Gresley, who was already carriage and wagon superintendent at Doncaster and who now combined both responsibilities. The outbreak of the First World War in 1914 ultimately resulted in fewer, but much heavier, and slower, passenger trains, which hitherto had not normally exceeded 400 tons. By 1916 these sometimes attained 500 tons, which now taxed the limited adhesive weight, officially 36 tons, but by then probably nearer the later-acknowledged 40 tons, of the 'large' 4-4-2s when getting these away from King's Cross and up the 1-in-107 gradient through Gasworks and Copenhagen tunnels. This called for piloting, which had been strictly forbidden for several years, over the 12½ miles to Potters Bar summit. In 1915, outline diagrams were prepared for four-cylinder 4-6-2s, and in 1918 for three-cylinder 2-6-2s, in both cases with narrow and wide firebox alternatives.

It was only immediately after the completion of the prototype large, three-cylinder mixed-traffic 2-6-0 No. 1000 in March 1920, that attention was then seriously focussed on designing a three-cylinder 4-6-2 express engine capable of handling trains of up to 600 tons. As completed, this would likewise carry 60 tons upon its coupled axles, representing in this respect a four-fold advance on 4-2-2 No. 1 of 50 years earlier. The new 4-6-2 was designated Class A1, as earlier had been the final 1895 series of Stirling 4-2-2s.

Approval to build two 4-6-2s was obtained in January 1921 and their assembly was well under way by the end of that year. The passing of the Railways Act, 1921 in August, foreshadowed the legal demise of the independent railway companies, including the GNR, through mutual amalgamation. The first 4-6-2, No. 1470 *Great Northern*, was only the second GNR locomotive to be named, making its debut in February 1922. It worked up to London for the first time a few weeks later, on 4 April.

Ten further 4-6-2s, Nos 1472–1481, were also quickly authorised thereafter, on 10 July 1922, from Doncaster Works under Engine Order (EO) 297. These were primarily intended for Anglo-Scottish express duties, of which the first to appear would quite unwittingly become a locomotive legend and the subject of this and several other books.

## DONCASTER WORKS

Britain's railway companies were highly unusual worldwide as to the extent to which they built their own locomotives in workshops. These had been established in the first instance simply to undertake routine repairs and not primarily locomotive manufacture. In terms of total steam locomotive production, at 2,224, between the mid-19th century and the late 1950s, Doncaster Works ranked fifth after its contemporaries at Crewe, Swindon, Derby and Darlington.

Opened in 1853 to undertake locomotive repairs, Doncaster Works, or The Plant as it was known locally, did not undertake new locomotive construction until 1867. Under British Railways ownership it completed its final steam locomotive just 90 years later, in October 1957, and its final diesel in early 1987. In addition to being built there, No. 4472 regularly underwent repairs at The Plant, in the course of which it spent an aggregate of almost exactly four years there during its conventional 40-year working life. It was reassembled in the Crimpsall erecting shop, which dated from 1900 and which was demolished in the spring of 2008.

**OPPOSITE An Ivatt 'large' 4-4-2, No. 1442 (built 1905), works a heavy express near Potters Bar, c1908. Following the (now-preserved) prototype No. 251 completed in December 1902, 90 generally similar engines were built at Doncaster between 1904 and 1910. These were all retired between 1944 and 1950.**

# CHAPTER FOUR

# No. 4472, its design and development, 1920–1960

A rather alarming photograph of LNER A3 class 4-6-2 No. 2598 *Blenheim* undergoing tests at Doncaster Works in September 1936, to determine the height of its centre of gravity. These actually confirmed the remarkable precision of the original calculations, while the slight differential between the respective centres of gravity when the boiler was full and empty can clearly be seen.

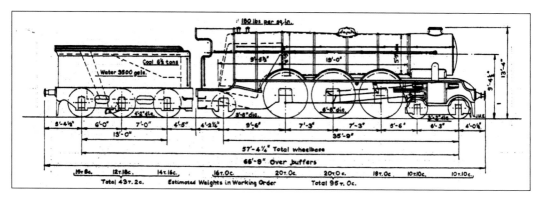

# Designing *Flying Scotsman*

No. 4472 was effectively conceived in Doncaster Works drawing office on Monday, 12 April 1920, with the drafting of a preliminary dimensioned diagram, officially designated 'sketch number 20-10' and entitled 'Proposed Passenger Engine 4-6-2 Type'. This was most probably executed by a leading draughtsman, Harry Broughton, who was later appointed chief locomotive draughtsman at Doncaster in April 1928. In its essentials the outline 4-6-2 was very close to GNR No. 1470 *Great Northern*, as it appeared almost two years later, except that superficially it featured a plain cab, had a straight running board, and was shown paired with a standard Ivatt six-wheeled tender.

**BELOW The historic entry in the Doncaster drawing office 'sketch' register, dated 12 April 1920, for the initial Gresley three-cylinder 4-6-2 diagram.**

Although given to making free-hand sketches, Nigel Gresley did not personally *design* in the drawing board sense the locomotives attributed to him; that was teamwork. Gresley would give directions, adjudicate on points of detail design, and approve the final drawings (which he only rarely

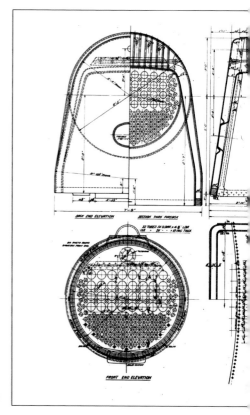

signed). He held the ultimate responsibility, and alone was empowered to authorise any subsequent modifications, however small.

Even the chief draughtsman was above actually producing drawings, being a key administrative, coordinating and very often highly influential figure. Never mentioned in the copious literature devoted to Nigel Gresley and his locomotives, William Elwess was listed as chief draughtsman at Doncaster in the 1906 edition of *The Railway Yearbook*. Remarkably little is known about Elwess other than that he was born *c*1867 on a farm at Wheatley, very close to Doncaster Works. Although his 'anonymous' notebooks survive in the National Railway Museum archives, disappointingly these make no reference whatever to the A1 4-6-2. Having overseen three major modifications to this, Elwess retired early in 1928, and died in Doncaster in 1946. His successor, Harry Broughton, retired in 1934 and died in 1947.

## Civil engineers' approval

Many of the drawings for the new 4-6-2 had been made by the end of 1920, before authorisation for its construction was given by the GNR Locomotive Committee. However, at an early stage, Gresley would have required provisional acceptance for his projected 4-6-2 from the chief (civil) engineers of the Great Northern and North Eastern railways, who would take into account its extreme dimensions, estimated weights and their distribution on track and bridges, etc. NER clearance was required for operation over its own metals beyond Shaftholme Junction, just north of Doncaster, and on to York.

## Notable features

The most remarkable feature of the 4-6-2 was its very large boiler, of which three later variants remained in production until 1961–62. By 1950, similar boilers were carried by a total of

**BELOW** General arrangement drawing of the original 4-6-2 boiler (180lb).

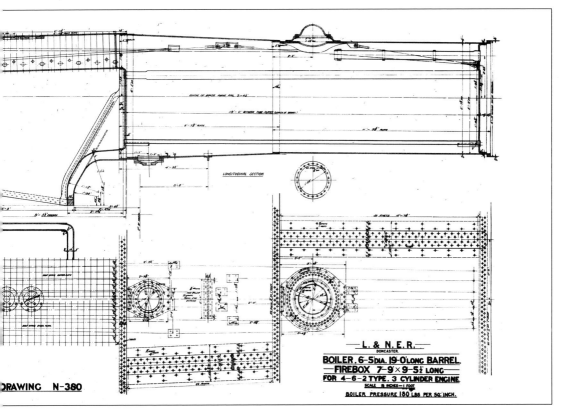

DRAWING N-380

L. & N. E. R.
DONCASTER
BOILER. 6-5"DIA. 19-0"LONG BARREL
FIREBOX 7-9 × 9-5½ LONG
FOR 4-6-2 TYPE. 3 CYLINDER ENGINE
SCALE 8 INCHES = 1 FOOT
BOILER PRESSURE 180 LBS PER SQ. INCH.

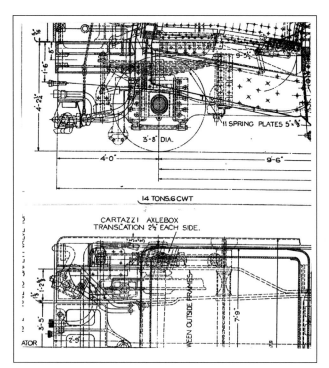

11 SPRING PLATES 5'·⅜

3'-8" DIA.

4'-0"

9'-6"

14 TONS.6 CWT

CARTAZZI AXLEBOX
TRANSLATION 2½" EACH SIDE.

**ABOVE** Drawing
showing Cortazzi rear
truck on A1/A3 class
4-6-2 indicating 2½in
side play.

When the first two Gresley 4-6-2s were authorised in January 1921, it was only then proposed to provide unusually large eight-wheel tenders, more in keeping with their size.

At least four features on the Great Northern Railway 4-6-2 could be traced directly back to 4-4-2 and 0-8-0 locomotives built around 20 years earlier by the Lancashire & Yorkshire Railway, from whence Nigel Gresley had come to Doncaster in 1905.

## Aesthetics

An enduring triumph of industrial design for such an unusually large locomotive for its time, the GNR 4-6-2 was remarkably well proportioned and extremely graceful. It was therefore particularly appropriate that the great majority of the 79 engines that would ultimately be built to this basic design, would be named after racehorses, including several Doncaster St Leger winners. This was a happy inspiration, attributed to Nigel Gresley personally, after he had received a directive from the LNER chairman, William Whitelaw in early 1925, to draft prospective names for the early production 4-6-2s.

The beautifully tapered painted sheet steel boiler clothing was complemented by the elegant reverse curves of the running board beneath – a throw back to the Stirling 4-2-2s. Another enhancing feature was the bold steam pipe casings above the cylinders, whose outward sweep anticipated the spread of the wide firebox aft.

nearly 400 powerful three-cylinder 2-6-2 and 4-6-2 locomotives, which shared a 22-ton axle load, putting them in the most restricted Route Availability (RA) 9 category.

An associated feature of all these locomotives was the Cortazzi (always officially spelt Cartazzi) truck with 2½in side play beneath the firebox. Francis Cortazzi had been the first works manager at Doncaster in the 1850s, but no patent has been traced for his truck, which Doncaster did not adopt until 70 years later for its 4-6-2s.

Gresley had not been happy with the Ivatt-style cab originally shown, and in early November 1921, three months after the assembly of Nos 1470 and 1471 had already begun, an unsolicited pencil sketch by apprentice Bert Spencer caught his eye for a more handsome double side-window cab, inspired by those of the Great Eastern Railway 4-6-0s. Deceptively, this was actually of no greater actual length than the initial cab proposal, and it was translated into a wooden scale model made by the pattern shop for Gresley's formal approval, before being finally adopted.

## Valve gear problems

The large, three-cylinder 2-6-0 made its debut in March 1920 and design work on the 4-6-2 clearly commenced almost immediately thereafter. Ironically, the 4-6-2's potential was initially compromised as a result of early experiences with the 2-6-0, which had been designed with long-lap (1½in)/long maximum travel (6⅝in) piston valves. Problems had been experienced almost immediately when coasting at high speeds in full gear with steam shut off, owing to over-travel of the middle piston valve by as much as 1in. This was due to insufficient rigidity in the mounting of the main pivot, and whiplash in the large '2-to-1' lever. The 4-6-2 had originally been intended to have the same lap and maximum valve travel as the 2-6-0, but in the light of this problem, maximum cut-off was limited to only 65 per cent rather than the possible 75 per cent. Maximum valve travel was correspondingly reduced to merely 4⁹⁄₁₆in. The lap was also trimmed to 1¼in, and after tests, other minor adjustments were later made to the valve characteristics, with lead and exhaust clearance increased from ⅛in to ³⁄₁₆in, and ¼in respectively.

From the outset, drivers tended to work the 4-6-2s at about 40 per cent cut-off in the long-established 4-4-2 tradition, which was also the 'earliest' cut-off able to provide a full port opening for exhaust. As a direct consequence, during their early years, the Gresley Pacifics proved to be neither as free running nor as economical in coal consumption as they could, or should have been.

### SHORT- VERSUS LONG-VALVE TRAVEL

In 1924, Bert Spencer, then a newly qualified apprentice, joined Nigel Gresley's personal staff as a technical assistant. He almost immediately identified the shortcomings of the 4-6-2 valve gear arrangements, and advocated Great Western-style long-lap/long valve travel (as indeed had originally been proposed) which would be more conducive to shorter cut-off working and improved fuel economy via the more prolonged continued expansion of steam after cut-off. Spencer's recommendations were initially dismissed by Gresley, but matters later moved on beyond even *his* control.

That February, the third 4-6-2, No. 1472, was renumbered 4472 and named *Flying Scotsman*, having been selected for display at the forthcoming British Empire Exhibition at Wembley. There, close by, was also to be exhibited the recently built yet distinctly smaller Great Western Railway four-cylinder 4-6-0, No. 4073 *Caerphilly Castle*, which the GWR proudly billed, purely on a starting tractive effort basis (31,625lb compared to the 29,835lb of the LNER 4-6-2), as 'the most powerful express locomotive in Britain'.

The GWR had also just withdrawn its legendary and solitary 4-6-2 *The Great Bear*, which despite its impressive bulk, had boasted only 27,797lb tractive effort. A further exhibition at Wembley took place the following year, at which No. 4472 was once again present, this

## POSSIBLE AMERICAN INFLUENCE?

In 1961, F.A.S. Brown suggested that the Gresley A1 4-6-2 might have been inspired by the celebrated Pennsylvania Railroad K4 class 4-6-2 of 1914, for which working drawings had been published in the British periodical *Engineering* two years later. Although Gresley never referred to any such influence, several years later he did freely acknowledge recent French practice regarding his P2 2-8-2s. Without question, however, the lightweight three-bar crosshead at least, already seen on the very recent Gresley three-cylinder 2-6-0s and 2-8-0s, had quite definitely evolved on the PRR more than 20 years earlier.

**ABOVE The three-bar slidebar and crosshead arrangement featured on the prototype Pennsylvania Railroad K4 class 4-6-2 No. 1737 built in 1914, shown here. (This was not perpetuated in the more than 400 subsequent production engines built 1917–28.)**

**BELOW Right-hand crosshead and slidebar arrangement on No. 4472.**

**RIGHT** The Doncaster drawing office valve gear model, on which the original short-travel valve gear for the A1 class 4-6-2 would have been designed, seen in use in the early 1950s. A similar analogue in the Darlington design office was later employed to plan the long-travel gear, using data covertly obtained from GWR 4-6-0 No. 4082 *Windsor Castle*.

time together with a different GWR 'Castle', No. 4079 *Pendennis Castle* nearby. Under G.J. Churchward, high boiler pressure (225lb) and long piston valve travel (of at least 6in) had already characterised GWR locomotives for 20 years, although Swindon itself in splendid isolation, would retain a peculiar indifference to adopting high-degree superheating for a further 20 years. Possibly because it enjoyed the luxury of Best Welsh steam coal, it rested content with moderate steam temperatures of only about 500°F, considered sufficient simply to minimise condensation in the cylinders.

### LNER V GWR LOCOMOTIVE EXCHANGE

Earlier, during April–May 1925, a locomotive exchange trial took place whereby a Gresley 4-6-2, No. 4474 (later named *Victor Wild*), ran on the GWR between London, Paddington and Plymouth, while a GWR 'Castle' 4-6-0, none other than No. 4079 *Pendennis Castle* in fact, worked between London, King's Cross and Doncaster. Just 40 years later, not long before his death in 1967, Bert Spencer recalled that even Gresley had learned of this forthcoming event via a newspaper report! He also confessed that he had never discovered how the exchange had actually come about and who had instigated it.

Both engines, when operating on the GWR and burning Welsh coal, were more economical than when they were working on the LNER using Yorkshire coal. The coal consumption of

the GWR 4-6-0 was lower than that of its LNER 4-6-2 counterpart when playing both at home (42.0 *v* 48.0lb/mile) and away (53.4 *v* 57.1lb/mile), which Gresley attributed to its significantly higher boiler pressure rather than its long-travel valves. He nevertheless promptly instructed that a 4-6-2 should be fitted up with long-travel valves hoping to prove that these were not justified. No. 4477 *Gay Crusader* was selected with its valve lap increased to 1⅝in and maximum travel extended to 5¾in. Re-entering traffic on 25 June this hasty improvisation unsurprisingly resulted in no improvements.

### 'INDUSTRIAL ESPIONAGE'

Only a few days later, two other GWR 'Castles' visited LNER territory to participate in the Stockton & Darlington Railway Centenary celebrations in the North East. Although their valve characteristics were already well known to Bert Spencer, during the course of these events, one night in early July 1925 at Darlington, the extremely inaccessible inside Walschaerts valve gear of one of them, No. 4082 *Windsor Castle*, was covertly measured up by a senior Darlington draughtsman. The dimensions were incorporated in the Darlington valve gear model, from which the valve events at 25 and 65 per cent cut-off were read off and tabulated.

No immediate action followed, but undeterred, Spencer continued to press for long-travel valves on the 4-6-2s throughout 1926. Early in the following year, Gresley finally relented and No. 2555 *Centenary* was modified in March 1927, closely based on the cribbed characteristics of GWR No. 4082. It could now readily be worked at only 15 per cent cut-off, which also still provided a free exhaust, and returned a coal consumption of only 39lb/mile when compared with 50lb by 'control' No. 2559 *The Tetrarch*.

Having also promptly ridden on *Centenary* up to London this time, Gresley was so impressed that he immediately ordered that all the 4-6-2s should be altered, which took place between November 1927 and May 1931 (No. 4472 in April 1928).

No longer necessary, the restricted maximum cut-off was nevertheless retained for many years, and not increased to 75 per cent (to give 6⅝in maximum travel) until after the Second World War, and only gradually between February 1946 and November 1958 (on No. 4472 in March 1948). A plate fixed in the cab instructed that 'when coasting with the regulator closed the reverser should be set at no more than 25 per cent cut-off'.

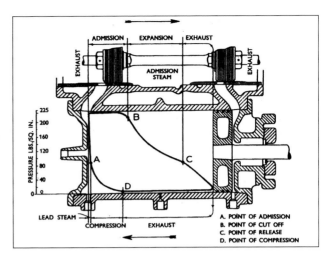

**ABOVE RIGHT** Diagram showing the steam expansion cycle in a locomotive cylinder.

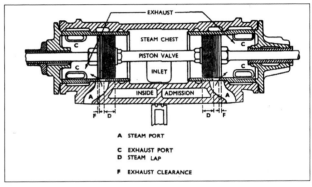

**RIGHT** Diagram explaining piston valve characteristics.

| VALVE CHARACTERISTICS AT 25 PER CENT CUT-OFF FOR LNER A1 CLASS 4-6-2 SHORT-TRAVEL, A1/A3 CLASS LONG-TRAVEL, AND BRITISH RAILWAYS CLASS 7MT 4-6-2 (FOR COMPARISON) | | | |
|---|---|---|---|
| | *LNER A1 4-6-2 short-valve travel (final setting, 1925)* | *LNER A1/A3 4-6-2 long-valve travel (1927)* | *BR Class 7 4-6-2 (1951)* |
| Cylinder dimensions, in | (3) 20 x 26 | (3) 20/19 x 26 | (2) 20 x 28 |
| Piston valve dia, in | 8 | 8 | 11 |
| Valve lap, in | (o) 1.25, (m) 1.31 | (o) 1.625, (m) 1.69 | 1.69 |
| Valve lead, in | 0.188 | 0.125 | 0.25 |
| Exhaust clearance*, in | 0.25 | nil | nil |
| Maximum valve travel, in | 4.56 (65% CO) | 5.75 (65% CO) 6.63 (75% CO) | 7.73 (78% CO) |
| At 25% cut-off: | | | |
| Mean port opening, in | (o) 0.30, (m) 0.31 | (o) 0.30, (m) 0.32 | 0.39 |
| Release | (o) 62.1%, (m) 62.4% | (o) 72.2%, (m) 72.1% | 69.8% |
| Compression | (o) 76.3%, (m) 77.6% | (o) 72.2%, (m) 72.1% | 69.3% |

(o) = outside cylinders, (m) = middle cylinder, % = of piston stroke, CO = cut-off
* also known as negative lap

# Evolution and development

## Alterations to suit the Scottish loading gauge

Standing at 13ft 3in from rail to chimney top the North Eastern 4-6-2s worked north of Berwick to Edinburgh, whereas the first 11 Gresley 4-6-2s at 13ft 4in could not. Their cabs would also have compromised the former North British loading gauge.

The 40 A1s ordered by the LNER from Doncaster Works and NBL in late 1923, Nos 2543–82, were patterned on No. 1481 in that their maximum height was reduced to 13ft 1in in order to permit them also to work on other former North British routes in Scotland. These radiated from Edinburgh, i.e. to Glasgow, Aberdeen and Carlisle. Most noticeable was the provision of shorter, 8½in tall chimney castings, and reduced cabs. Nos 1470–80 (as Nos 4470–80) were altered to conform with their successors between February 1928 and September 1933 (No. 4472 in April 1928).

Also, for platform clearance reasons, particularly within Newcastle Central station, the bottom corners of the squared-off front buffer beams of these early engines were altered by the deletion of small inset quadrants (on No. 4472 in April 1927).

## A1 expanding spheres of operation

The 4-6-2s were not permitted to work on and into Leeds over the former GNR route until 1930 because of weight limits on the single-span steel bridge over the River Calder at Wakefield. However, they appeared sooner in Leeds from the north, coming over the former NER route via Ripon and Harrogate. The southern leg of the East Coast London–Edinburgh through express passenger workings had traditionally been serviced by the King's Cross and Grantham locomotive sheds. No. 1471, when temporarily on loan to King's Cross depot during the late summer of 1922, had already worked regularly as far north as Grantham on the then-styled 10am 'Special Scotch Express' ex-King's Cross, which only a year or so later was officially re-titled the 'Flying Scotsman' by the LNER.

By late 1923, representatives from the first 12 Great Northern-type 4-6-2s were beginning to appear at York, and during the following summer, King's Cross 4-6-2s began to work the 'Flying Scotsman' regularly thus far. Mid-1927 saw them extending further north to Newcastle. A1 No. 1481, with cut-down cab and boiler mountings, had already worked into Edinburgh from Newcastle as early as November 1923. Edinburgh Haymarket shed received its first allocation of similarly modified brand-new A1s in August 1924, and Gateshead and Heaton (Newcastle) sheds in September and October respectively. By late 1928 operations extended from Edinburgh to Carlisle and Aberdeen, while ten years later, in 1938, A1s displaced by new A4s, also began to operate over the former Great Central Main Line between London Marylebone and Manchester.

## 'Super Pacifics'

In March 1927, five new 4-6-2 boilers with their working pressure increased from 180lb to 220lb per sq in, and equipped with enlarged superheaters inspired by German practice, were ordered. The first was fitted to No. 4480 *Enterprise* in July 1927, and the remainder up to May 1928. Four of these conversions initially retained their original 20in cylinder bore in conjunction with the higher 220lb pressure, which besides making for potentially very 'slippery' engines, their relative cylinder demand and boiler capacity also fell badly out of kilter. No further such conversions were undertaken until 1939. When No. 4472 received long-travel valves in April 1928, it was also fitted with the last new 180lb boiler.

---

## LONDON–EDINBURGH NON-STOP

On 11 July 1927, No. 4475 *Flying Fox* made an unusually long non-stop run from King's Cross to Newcastle, 268½ miles in 5½hr, with Driver Pibworth. As at this time the engine was still fitted with short-travel valves, at 50lb per mile this would have entailed his fireman shoveling about 6 tons of coal en route.

The substantial fuel savings wrought by the adoption of long-travel valves resulted in greatly increased endurance, which therefore offered the prospect of longer through runs. Even London–Edinburgh non-stop would be possible by providing special corridor tenders permitting crew changes mid-way while on the move. The non-stop 'Flying Scotsman' service was inaugurated on 1 May 1928. (See Chapter 6.)

Ten new 4-6-2s were authorised in August 1927, which would combine long-travel valves with 220lb boilers, and these were officially classed A3 (as already the ex-North Eastern Railway 4-6-2s were A2). Popularly dubbed 'Super Pacifics', they began to appear in August 1928 and were also fitted with modernised non-corridor versions of the new corridor tenders. The newly built A3s had 19in diameter cylinders with which larger 9in diameter piston valves would have been possible, but so that the same cylinder castings could serve both classes, liners were inserted to reduce the cylinder bore from 20in to 19in on the A3s. The latter could always be readily distinguished by the square 'blisters' or cover plates on the sides of the smokebox, which were necessary to accommodate the larger and wider superheater header.

## Matters of some weight

Although the 220lb pressure boilers reportedly weighed 2½ tons more than the 180lb units, the *official* total weight increase of the A3 over the A1 was nevertheless almost 4 tons. The adhesive weight increase was no less than 6 tons, from 60 to 66 tons (the coupled axle load had increased from 20 to 22 tons) preserving the A1 adhesion factor of 4.5.

A1 No. 4481 *St Simon* when weighed in December 1926 returned 97½ tons, a surprising 5 ton increase on the 92½ tons officially stated for the almost identical *Great Northern* in 1922. (No. 4472 as an A3 scaled 97 tons when weighed in 1967.)

The first five 4-6-2s built by the North British Locomotive Co. were stated to weigh 93½ tons in working order. Their stipulated weighing conditions included 2,100 gallons (9.4 tons) of cold water in the boiler at 'half (gauge) glass', 6½cwt of coal on the grate, 4½cwt of sand, and the standard allowance of 3cwt for the engine crew.

## The benefits of increased boiler pressure

Higher steam pressure, via the greater pressure drop, made for increased cylinder efficiency, and hence lower fuel consumption. Other things being equal, tests showed respective cylinder efficiencies of 13.2 and 14.3 per cent for 180lb

and 220lb pressure, a relative enhancement of 8 per cent, which was reflected in fuel economies of 5 per cent by the A3, at 35.4 *v* 38.3lb/mile. Also in purely practical terms, when the maximum cut-off was still restricted to 65 per cent, the 10 per cent increase in tractive effort of the A3s proved advantageous in helping to re-start heavy up trains when halted by signals on the long 1-in-96 Cockburnspath Bank north of Berwick, or when working over the demanding Edinburgh–Carlisle route. In Scotland, on all routes concerned, the A3s were permitted to take a 50-ton greater maximum loading than the A1s, but A1s and A3s were employed indiscriminately on the non-stop 'Flying Scotsman' workings.

After the first batch of A3s had been built, when more similar engines were due to follow, a report was produced in January 1930 which audited the benefits or otherwise of the first experimental high-pressure conversions by making comparisons with the original, standard 180lb A1. It noted only boiler repair costs were significantly increased due to the greater associated stresses and higher working temperatures. The report rather pessimistically concluded that 'it is evident that any economies likely to accrue in coal and water consumption on the 220lb pressure boilers will be more than neutralised by the increased cost of maintaining these boilers.' It followed that the higher pressure could also result in shorter boiler life, and so amounted to a trade-off of higher repair/renewal costs versus reductions in fuel bills via increased cylinder efficiency. Nevertheless, 250lb was adopted on the A4 4-6-2s in 1935, and only four years later Gresley was initially considering even 300lb boiler pressure for his 'swan song' V4 2-6-2.

## The quest for hotter steam

In the original 180lb A1 boilers the maximum steam temperatures achieved by their 32-element superheaters had proved to be much lower than anticipated, scarcely reaching 600°F, against the contemporary target of 650°F. Above that it was generally believed there could be piston valve and cylinder lubrication problems as a result of carbonisation.

Trials in 1925 with the American E-type superheater proved unsuccessful. Even with the later 220lb A3 boiler an increase in the number of superheater elements from 32 to 43, and of their collective heating surface by 34 per cent, resulted in an increase of merely 30°F. Half of this was simply attributable to the further elevated boiling point of water due to the higher pressure. The same tube and flue layout on the post-war Peppercorn A1 and A2 4-6-2s yielded similar maximum steam temperatures of only 615–625°F, yet paradoxically, produced a peak of 735°F on a V2 2-6-2 on test in 1952.

## Post-war A3 performance

With the confluence of long-travel valves and 220lb boiler pressure, the effective format of the basic A3 class was established, which would then remain virtually unchanged for 30 years. At the very end of this period, in early April 1957, purely for practical operational purposes, dynamometer car trials were conducted with A3s on normal service trains of 315–455 tons (nominal).

The 'Tyne-Tees Pullman' and 'Talisman' were operated at average speeds of around 60mph between London King's Cross and Newcastle, and Edinburgh, on which the more recent A4 and Peppercorn A1 4-6-2s were also employed. The primary object was to establish the steam requirement for train heating as this could determine whether or not an extra coach could be attached to the formations during the winter months. The results also gave some insight into the capabilities, and limitations, of the Gresley A3s in their traditional form, and on the very eve of their final modification in which would prove to be their eleventh hour.

Limited by the physical capability of the fireman, the maximum continuous output at the drawbar on level track (therefore overcoming only frictional and aerodynamic train resistance) was established to peak over the speed range of 40–45mph at 1,235dbhp. This reduced to 1,160dbhp at 60mph, which would theoretically pull about 750 tons of British Railways 1950s Mark 1 passenger stock. However, on a rising gradient a locomotive is also required to *lift* both its own weight and that of its train, thereby necessitating significant reductions in speed and/or train loading.

Following the construction of 49 new Peppercorn A1 4-6-2s during 1948–49, the 78 by now ageing A3s accounted for almost half of the express 4-6-2s with 6ft 8in diameter coupled wheels employed on the ECML. They were now perceived as being the 'weakest link', hence their specific selection for the trials.

**BELOW** No. 4472 heads the pre-war 'Harrogate Pullman' through Doncaster.

Cook soon instituted pressed-in brasses in outside connecting rod big ends, a feature adopted at Swindon and derived from the French four-cylinder compound 4-4-2s imported by G.J. Churchward from France during 1903–05 for evaluation on the GWR.

Previously, during his long tenure as works manager at Swindon during the last ten years of the independent GWR, and inspired by recent similar developments in Germany, Cook had developed sophisticated optical techniques to align cylinders and frames extremely accurately during locomotive construction and repair, which he later instituted at Doncaster Works. He described the procedures to the Institution of Locomotive Engineers in his Presidential Address entitled 'The Steam Locomotive: a Machine of Precision' in September 1955, just as this much-loved human creation was rapidly falling from favour, almost the world over.

## The end of the line

The *gradual* phasing out of steam traction initially envisaged by British Railways was very quickly abandoned, and within only ten years the steam era was already nearly over. Although the first A3 had been withdrawn in December 1959, many others nevertheless continued to undergo heavy repairs and receive full repaints at Doncaster Works over the next three years.

The 22 'production' 'Deltics' had been ordered in 1958, officially to replace 55 A3 4-6-2s, and soon after their delayed first appearance three years later, the systematic retirement of the 4-6-2s commenced in September 1961 and peaked during 1963, when 33 A3s were retired. The last to remain in traffic was the 41-year-old BR No. 60052 *Prince Palatine,* which had also been one of the last to receive a heavy repair at Doncaster Works, in October 1962. This was condemned on shed in Edinburgh in mid-January 1966, and apart from the double chimney and smoke deflectors, its general appearance differed remarkably little from the pioneer *Great Northern* in 1922.

The longevity of the Great Northern Railway three-cylinder 4-6-2, originally drafted back in 1920, was indeed a testimony to the vision of Nigel Gresley interpreted by William Elwess and his Doncaster drawing office team. *Possibly* influenced by current American practice at its conception, it had later benefited from French and German locomotive technology, but remained essentially British to the end.

Although no fewer than six out of the later 35 Class A4 4-6-2s built have been preserved, of the 78 A1/A3s only No. 4472 has survived.

### Tornado

No. 60163 *Tornado* was finally completed with considerable enterprise in 2008 as the 50th and final member of the Peppercorn Class A1 4-6-2,

**ABOVE** New Peppercorn A1 4-6-2 No. 60163 *Tornado,* built in 2008, outside the National Railway Museum, York. See *The A1 Locomotive Trust Tornado Owners' Workshop Manual* by Geoff Smith, from Haynes Publishing, 2011.

built 1948–49 (BR Nos 60114–62). This can be regarded as the ultimate development of the original *Gresley* A1 Pacific of 1922. There were, however, important design differences which reflected changing conditions. On the Peppercorn A1, the middle cylinder was moved well forward driving on to the leading coupled axle, being provided with its own independent set of Walschaerts gear. This permitted the provision of larger diameter (10in) piston valves. Tube length was reduced from 19ft to 17ft and the grate area increased from 41¼sq ft to 50sq ft.

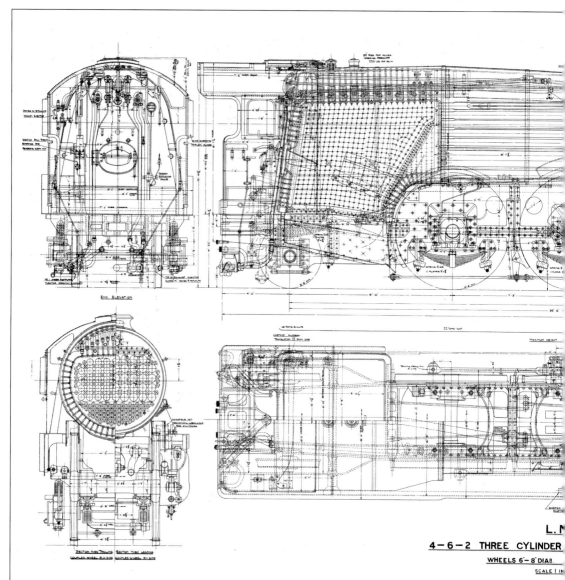

L. N

**4 – 6 – 2  THREE  CYLINDER**

WHEELS 6'- 8" DIAB.

SCALE I IN

Like Nos 60153–7, No. 60163 is equipped with the luxury of roller bearings, rather than traditional plain bearings throughout. The boiler (which was made in Germany) is also fully welded, although the originals had been of conventional riveted construction.

**BELOW** Sectional general arrangement drawing of LNER Class A3 4-6-2, Q-96. Strictly speaking this applies to the earlier 1928–30 builds, as the final 1934–35 batch were turned out with 'banjo' dome (94A) boilers, which were later also fitted to most of the class.

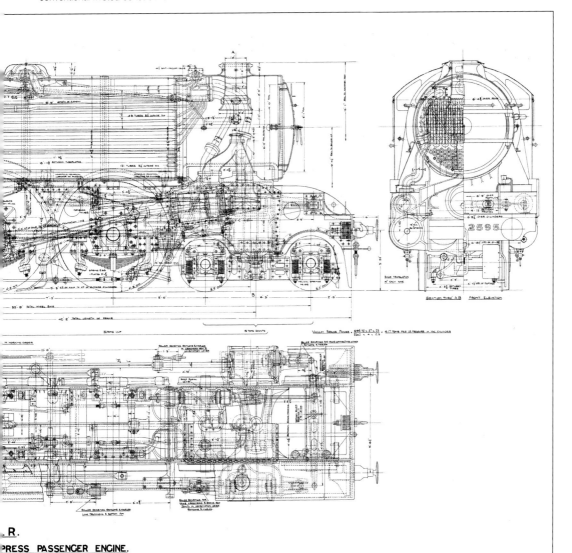

.R.
PRESS PASSENGER ENGINE.
CYLINDERS 19"x26" A.3.CLASS
1 FOOT

Dᴳ Nº 9168 B.

DRAWING Nº Q-96 N

GNR No. 1470's frames set up, in November 1921.
The middle cylinder and smokebox saddle have
already been installed, but the outside cylinders
have yet to be mounted.

CHAPTER FIVE

# Construction

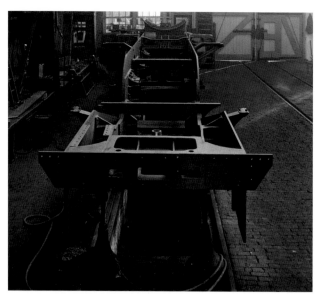

# Frames

The main frames provided the foundation of a steam locomotive, and in British practice these were invariably cut from steel plate. The frames provide a good example of the evolution of a major locomotive component on the drawing board and how this dovetailed into the construction process.

The general outline was first defined by a frame plate 'sketch', numbered 20-67 and dated

## SPECIFICATION (1923)

*FRAMES*

- To be of Acid Open Hearth Steel only, to LNER Specification No. 24.
- Each frame must be made in one length without a weld.
- All holes to be marked from one template, and drilled and reamered to the exact sizes given.
- When frames are bolted together with cross-stays and cylinders, the accuracy of the work must be carefully tested by diagonal, longitudinal, and transverse measurements.
- Frames to be finished with a good, smooth, and true surface and free from cross windings, the axlebox slides being square with the engine in all directions.

15 October 1920, which was produced for the Stores Department for ordering purposes. The 1⅛in thick frame plates for the 4-6-2 were ordered from The Leeds Forge Co. as rectangles measuring 42ft 4in long by 4ft 8in wide, weighing approximately 4 tons apiece as delivered. The *precise* frame profile was defined by a detailed frame outline drawing, O-134, completed in January 1921 (which interestingly, carried the tiny initials HB, i.e. Harry Broughton). From this a full-size wooden template could be made, from which the frame profile would be marked out directly on to the metal, about ¼in oversize, which would be cut out by a gas flame.

Pairs of frame plates, or even several pairs of frames simultaneously, would then be profiled or 'slotted' precisely on a special frame-slotting machine. The frame arrangement drawing, O-137, dated July 1921, further showed the precise location of *all* rivet and bolt holes which then had to be drilled in the frame plates.

The two frame plates were held together by stretcher castings which kept the two plates the traditional 4ft 1½in apart (as for a narrow firebox), for the greater part of their length, before splaying into four plates from a little behind the rear coupled wheels, in order to support the wide firebox. After the boiler, the frames accounted for quite a significant proportion of the total weight of the locomotive, and their thickness tended to be a trade-off of weight versus performance. Ideally this should have been 1¼in, as was later employed in the British Railways Standard 4-6-2s, and would only have increased the total engine weight by scarcely 1 ton. Such would both have reduced the incidence of cracks in the frames and the need for their consequent repair and eventual replacement.

In fact, the frame performance of the non-streamlined Gresley 4-6-2s was particularly poor. One A3 built in 1930 required new frames after only three years, although *Great Northern*'s original frames (illustrated) lasted 11 years. Seemingly, all the A1/A3s' frames were eventually replaced, at least ahead of the firebox, at some point. In No. 4472's case, although unrecorded, this was most probably during 1933–34, when all three cylinders were changed twice.

Doncaster's relaxed philosophy was by no means unusual; a former 1930s apprentice

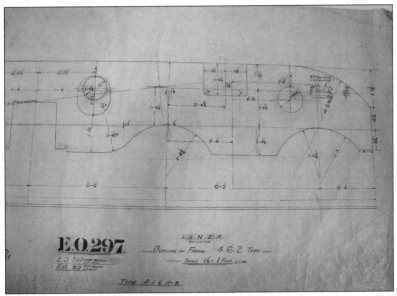

**LEFT** Detail from Doncaster working drawing No. O-134 '4-6-2 frame outline' for the front end of the 4-6-2 main frames, dated January 1921.

**BELOW** Detail from Doncaster drawing No. O-137, '4-6-2 frame arrangement'. Dated July 1921, 93 years later this portion of the working drawing was used to produce the new front frame section for No. 4472 in 2014.

Both documents are Indian ink on linen tracings drawn to a scale of 1½in to the foot.

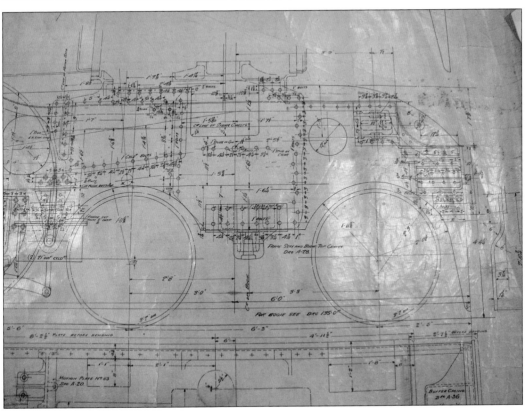

there, Eric Beevor, recalled in *Steam Was My Calling* (Ian Allan, 1977) that 'not only were boilers and tenders freely interchanged between locomotives of a particular class – and sometimes from one class to another – but sets of wheels, cylinder blocks and even the basic 'chassis' (i.e. frames) of a locomotive might be changed.' Metallurgical analyses of No. 4472's frame plates in 2014 unexpectedly revealed these to differ from each other, and so possibly to vary as to their provenance.

Major frame fractures on the A1/A3 4-6-2s would for ever remain an on-going problem. As BR No. 60112 *St Simon*, like No. 4472 also of 1923 build, rather surprisingly had its frames completely replaced ahead of the firebox as late as October 1962, although by this time several later-built A3s had already been retired.

# Bogie

A leading four-wheeled bogie was highly desirable in a powerful express locomotive, particularly in order to guide it on curved track at high speed.

The original bogie design closely followed that of the earlier Ivatt 4-4-2s with swing links, and perpetuating their 6ft 3in wheelbase. This did not entirely suit the Gresley 4-6-2s and after a few years their bogies were modified on the lines of that later designed afresh (but with 6ft 6in wheelbase) at Darlington for the D49 class 4-4-0, which carried a 7 ton greater loading. Its total side play was increased from 3½in to 4in (which from 1931 was spring controlled after the removal of the swing links) and the journals were lengthened from 9in to 11in, together with the provision of stiffer coil bearing springs.

The fracture of the right-hand bogie frame plate on the solitary (and by then rebuilt) Gresley 4-6-4 as BR No. 60700, had resulted in its

---

**SPECIFICATION** (1923)

*BOGIE*

- Frames, material and workmanship, to be of the same quality in every respect as the Engine Frames.

---

**LEFT** No. 4472's bogie at Bury, August 2012.

spectacular derailment at Peterborough on 1 September 1955, fortunately without serious consequences. Therefore the thickness of the bogie frames on all the ex-LNER 4-6-2s was thereafter increased from 1⅛in to 1¼in as and when these required renewal. (The bogie frames on No. 4472 as preserved are 1¼in thick).

# Cylinders

The manufacture of locomotive cylinders was a particularly complex and extremely skilled three-dimensional operation, involving the prior fabrication of intricate wooden patterns, which in itself could occupy up to three months. The process was described in rare and considerable detail with respect to the immediately preceding Great Northern Railway three-cylinder 2-6-0, in an illustrated article in *The Railway Engineer* for April 1920. The crucial part is reproduced herewith:

*The drawings supplied to the pattern shop from the drawing office are set out on a scale of 3 inches to the foot and fully dimensioned. The dimensions are read off and sections of the cylinders reproduced, in full-size, by the pattern makers on large boards. When these sections or parts of the cylinders and steam chests have been set out full-size on the boards, the foreman pattern maker decides where the partings or sectionising (sic) of the patterns shall occur, such divisions being absolutely necessary to allow of the patterns being withdrawn from the sand when the mould has been completed … The pattern makers employ a set of rules for measuring the pattern to allow for the contraction of the metal (molten at 2,000 degrees Fahrenheit). These rules are suitably marked for different coefficients of expansion. The drawings are marked with the aid of these rules, and this saves adding and calculating, shrinkage in the casting being automatically allowed for (⅛in per foot for cast iron). The allowance for machining, in this case ³⁄₁₆in as an average, is also provided for by the pattern maker.*

**RIGHT** Moulds and cores for 2-6-0 cylinders, Doncaster Works, 1920.

**ABOVE** A close-up of the right-hand outside cylinder casting mounted on No. 1470's frame. As with the boiler, the cylinders were provided with insulation which was contained within sheet metal clothing, later painted, which was secured directly to the casting.

**BELOW** GNR 2-6-0 outside cylinder patterns, Doncaster Works, 1920.

**SPECIFICATION** (1923)

*CYLINDERS*

- To be perfectly free from honeycomb, and sound throughout, and in accordance with LNER Specification No. 14. To be of 'Warners' Cylinder metal, to specification which will be supplied, as hard as can be bored and planed.
- Steam Chest Liners to be of cylinder metal, accurately machined and shrunk in the cylinders as shown on the drawing.
- The cylinders to be tested by hydraulic pressure of 250lb per square inch.
- All joints to be scraped so as to be perfectly steam-tight, to be lagged with asbestos, and fitted with drain cocks, as shown on the drawing, securely bolted to the frames with turned bolts driven tight.
- The joint between the outside cylinders and saddle casting to be a faced joint.
- All gland, steam chest cover, and cylinder cover Nuts, to be case-hardened.

**ABOVE** The left-hand side view of a 4-6-2 middle cylinder casting with a crack highlighted, at Doncaster Works in 1931. These castings usefully doubled as massive main frame cross-members, reinforcing the inherently rather weak frames at a potentially vulnerable point above the cut-out for the rear bogie wheels. This was close to the outside cylinders, which would have set up considerable stresses in themselves.

**RIGHT** The middle cylinder removed from No. 4472 during its 1996–99 overhaul (rear view).

Although mounted horizontally, the two outside cylinders on the 4-6-2 were 'handed' and therefore although not interchangeable with each other, were cast using the same wooden pattern. In their finished state each weighed 1½ tons, the middle cylinder casting slightly more. Cast iron is a relatively soft and brittle metal, therefore subject to wear and prone to cracking, and during the first 15 years alone, each of the three cylinders on No. 4472 was replaced four to five times, either by new or reconditioned second-hand castings. No records are available after 1938.

| NO. 4472'S CYLINDER REPLACEMENTS, 1923–38 | | | |
|---|---|---|---|
| Date | Left-hand cylinder | Middle cylinder | Right-hand cylinder |
| February 1923 | New engine, 20in | New engine, 20in | New engine, 20in |
| April 1928 | New cylinder, 20in | Re-bored to 20¼in | New cylinder, 20in |
| June 1929 | Re-bored and linered to 19¾in | Re-bored and linered to 19¾in | Re-bored and linered to 19¾in |
| March 1930 | Re-bored to 19¹³⁄₁₆in | New cylinder, 19¾in | Re-bored to 19¹³⁄₁₆in |
| April 1933 | Second-hand cylinder*, 19¾in | Second-hand cylinder*, 19¾in | Second-hand cylinder*, 19¾in |
| May 1934 | New cylinder | Second-hand cylinder | Second-hand cylinder |
| June 1936 | - | - | New cylinder, 19¾in |
| July 1938 | - | New cylinder, 19¾in | - |

* Ex-No. 2564 *Knight of the Thistle*

## Why *three* cylinders?

All 400 British 4-6-2s built prior to 1951 had either three or four cylinders. This was simply because sufficient power could not be developed by only two (outside) cylinders. Their diameter was restricted by the British loading gauge so as not to exceed a maximum overall width, over clothing, of barely 9ft. (In the case of No. 4472 this was 8ft 10in.) For the Gresley A1 with three 20in cylinders, the diameter of two cylinders of equivalent volume would have had to be made 24½in, or 23¼in in the A3, whereas the limit within the British loading gauge was usually 21in. Also, the lower hammer blow of three- and four-cylinder locomotives than those with only two cylinders meant that these could be accepted with a heavier axle load, e.g. 22½ v 20 tons (see Balancing on page 54).

By 1925, Nigel Gresley had developed a distinct preference for three-cylinder locomotives in general, crediting them with lower coal consumption through shorter cut-offs, and an increased mileage between repairs, when compared with a similar (but cheaper and less complicated) two-cylinder engine.

Gresley had insisted that the three cylinders in his 4-6-2s should each drive the *centre* coupled axle, in order to permit the leading coupled axle greater lateral flexibility. This, however, required the middle cylinder to be set between the frames about 3ft behind the outside cylinders, raised and inclined at 1 in 8, or an angle of 7°, and acting through a correspondingly shorter connecting rod.

The outside cylinders were horizontal, and following advice in early 1919 from Harold Holcroft, then on the South Eastern & Chatham Railway, the cylinder inclination differential was compensated for by advancing the inside crank angle by 7° to produce a phasing of 113° + 120° + 127°, rather than a strictly uniform 120° + 120° + 120°.

An inherent and indeed attractive characteristic of three-cylinder locomotives is six exhaust beats per revolution of the coupled wheels, compared with only four beats by either two- or four-cylinder engines with their 90° crank settings. Apart from delivering a more even torque this makes for a more uniform draught on the fire and also results in a softer but noticeably more rapid sounding discharge.

**LEFT Rear close-up of No. 4472's left-hand cylinder showing cylinder cocks and drain pipes.**

## WATER IN CYLINDERS

The presence of water in locomotive cylinders is highly undesirable because it is incompressible and it can lead to fractures in the cylinder walls. When standing, a small amount of water can form simply as a result of condensation, and this is expelled by routinely opening the cylinder cocks when starting away. Under certain circumstances while in motion, water can also be carried direct from the boiler to the cylinders, termed 'priming', and so, to allow for this compression, which could result in cracked cylinders, although not initially provided on the first Gresley 4-6-2s, relief valves were later fitted at low points at both ends of each cylinder.

**BELOW Front of left-hand cylinder showing the smaller compression relief valve.**

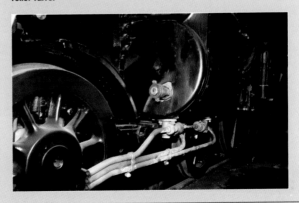

# Valve gear

A trademark feature of the great majority of
Gresley's locomotives was the employment of
three cylinders, in which the piston valves of the
outside pair employed direct Walschaerts valve
gear, and a simple system of floating levers
linked to this, imparted a derived or conjugated
motion to the piston valve of the middle cylinder.
While avoiding the more difficult alternative of
accommodating a set of inaccessible valve
gear between the frames, it was an imperfect
and distinctly controversial solution for a variety
of reasons. It dogged Gresley's three-cylinder
locomotives in general, and his 4-6-2s in
particular, in regular operation over a period of
more than 40 years. It also endowed the 4-6-2s
with their distinctive sound when in action.

Nigel Gresley had been granted a patent
for two forms of derived valve gear in October
1916 (BP 15,769/1916). The first and more
complicated form was uniquely applied to GNR
2-8-0 No. 461 built in 1918, which had its
three cylinders uniformly and steeply inclined
at 1 in 8, but his preamble to the specification
for the simpler version, which he subsequently
extensively employed, stated:

*'This invention relates to valve gear for
locomotive or other reversible steam engines
having three cylinders the piston rods of which
are connected with cranks set at suitable
angles, for instance 120 degrees, to each other.*

*'The invention consists principally in
providing connections between the valve
spindles or rods of two outer cylinders
(which may be situated outside the engine
frame in the case of a locomotive) and the*
*valve spindle or rod of an intermediate or
central cylinder whereby the movement given
to the valve spindles of the outside cylinders
effect the requisite movements of the valve
spindle of the intermediate cylinder so that
no independent or separate valve gear,
such, for instance as that usually employed
for each valve of a two-cylinder engine, is
required for actuating the distribution valve of
the intermediate or central cylinder.*

---

## SPECIFICATION (1923)

*VALVE GEAR*

- All the Valve Gear, except the large Motion
  Lever and Reversing Shaft Arms, to be made
  of Mild Steel to LNER specification No. 11,
  Class 'A'.
- The motion rods to be in one piece without
  welds, got up bright, and well finished
  throughout.
- All working parts must be well case-
  hardened ⅛in deep, and oil cups &c
  provided as shown on the drawing.
- Large Motion Lever to be Nickel Chrome
  Steel, heat-treated, to specification
  attached.
- Machining before heat-treatment &c to be
  the same as for Connecting and Coupling
  Rods.
- 'Hoffman' Roller and Ball Bearings to be
  fitted to centre pivots of large motion lever
  and equal motion lever.
- The eccentric rod bearing and reversing
  screw to be fitted with 'Skefco' Ball Bearings
  as shown on the drawings. All the Ball and
  Roller Bearings must be fitted with great
  accuracy throughout.

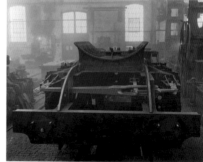

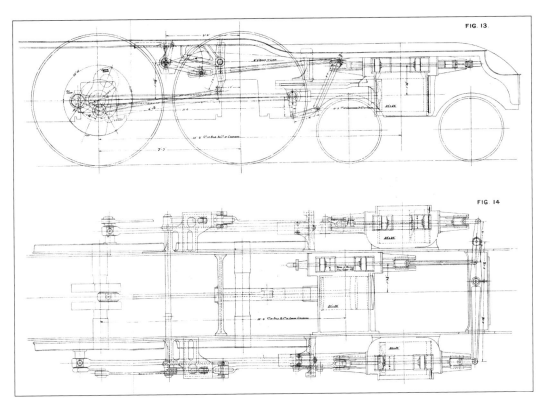

FIG. 13.

FIG. 14

'...for the purpose above-mentioned, I employ two rods or levers of unequal length and arranged transversely of the engine, preferably so as to work in horizontal planes one above the other, these rods or levers being operatively connected one with the other and with the valve spindles of the three cylinders. Each of these rods is pivotally connected, for instance by pins and links, at or adjacent to its outer end to the valve spindle of the adjacent outer cylinder, and the longer rod or lever pivots, intermediate of its length, on a fixed point at or near the centre line of the engine. The shorter rod or lever pivots intermediate its length on or near that end of the longer lever opposite to the end connected (as above-mentioned) to the valve spindle of one of the outer cylinders, the said short rod or lever thus being a floating lever. The valve spindle of the intermediate or central cylinder is pivotally connected, for instance by a link and pins, to the inner end of the aforesaid shorter rod or lever.'

**ABOVE** Simple valve gear arrangement drawing for the Gresley Class A1 4-6-2.

The idea was by no means original and almost certainly owed much to a very similar mechanism which had been patented in 1909 by Harold Holcroft, when he had been a GWR draughtsman at Swindon, although

**BELOW** The three piston valve spindles, each with their twin, 8in diameter valve heads.

this patent had since lapsed. However, despite
its simplicity, Gresley's valve gear had two
serious shortcomings:

1. Its geometry could seriously restrict the
   diameter of the piston valves, which ideally
   should be at least half the cylinder diameter.
2. The middle valve even at low speeds could
   'inherit' erratic valve events due to wear in
   the pin joints of the gear, which was further
   compounded by the thermal expansion
   of the outside valve spindles if this was
   mounted in front of the cylinders.

ABOVE Sectioned LNER 4-6-2 middle big end 'stink bomb'.

BELOW Union Pacific Railroad three-cylinder 4-12-2 locomotive with
Gresley valve gear. *(Union Pacific Railroad)*

## 'Stink bombs'

The erratic characteristics of the derived valve
gear could lead to an uneven distribution
of work between the three cylinders, of
which the middle cylinder tended to work
harder at high speeds. This could cause the
middle big end to run hot, and in extreme
instances to lose its white metal bearing (as
indeed occurred at the end of A4 No. 4468
*Mallard*'s record-breaking high-speed run
on 3 July 1938). Some A4s were provided
with so-called 'stink bombs' containing amyl
acetate ('pear drops') or aniseed, inserted
in their middle big ends as a precautionary
measure during 1937–39. Designed to
'detonate' at 160°F, i.e. well below the melting
point of the antimony-based white metal, they
emitted an unmistakeable smell detectable
on the footplate by way of a warning to the
engine crew of incipient heating occurring in
the middle big end bearing.

The Gresley-derived valve gear briefly
received a wide application almost worldwide
during the late 1920s/early 1930s, in North
and South America, Nigeria and South
Africa, Japan, Australia and New Zealand,
particularly on 2-8-2s, 4-6-2s and 4-8-2s.
Undoubtedly its most spectacular application
was on the 88 Union Pacific Railroad three-
cylinder 4-12-2s built by the American
Locomotive Co. during 1926–31, upon which it
was shamelessly exposed directly beneath the
front of the smokebox.

The various functional shortcomings of the
derived valve gear were further exacerbated by
the reduced maintenance conditions brought
about by the Second World War through the
shortage of manpower. Gresley's successor
as CME, Edward Thompson, sought an
independent appraisal of the '2-to-1' gear as
it stood, from William Stanier, his opposite
number on the LMS, who assigned E.S. Cox
to the task. Cox's highly analytical 1,400-word
report, dated 8 June 1942, arrived at the
following conclusions and recommendations:

1 *The '2 to 1' valve gear although theoretically
correct is, in practice, incapable of being
made into a sound mechanical job, and
rapid wear of the pins, and incorrect steam
distribution, are the inevitable results of its*

use. In view of its inherent defects and the discontinuance of its use throughout the world, a good case can be made for not perpetuating it in any future design.

2 It is certain that with this arrangement of valve gear it will be necessary to give the engines a frequent overhaul in the Shops and even then it is not possible to eliminate the effect of lost motion due to running clearance required in the pin joints and the effects of expansion of the outside valve spindles on the inside valve.

In private correspondence in 1964, Harold Holcroft dismissed Cox's report as having been 'very negative'. In his memoirs published two years earlier (*Locomotive Adventure*, Ian Allan, 1962) when recalling his collaboration with Gresley on the valve gear, he modestly observed 'although my part in it was forgotten after a time I was quite content with the quiet satisfaction of watching its progress over the years'.

## Connecting rods

The adoption of high-tensile nickel chrome steel halved the weight of the three connecting rods and coupling rods, at 3cwt each, and 4cwt per side respectively, compared with conventional mild steel, greatly facilitating the balancing arrangements.

### SPECIFICATION (1923)

*CONNECTING AND COUPLING RODS*
- Connecting and Coupling Rods to be made of Nickel Chrome Steel, heat-treated, to specification.
- To be fitted with bronze bearings with white metal pockets where shown on the drawings.
- All rods to be first machined with a surface allowance of ¼in for finishing, and then sent to the makers for heat treatment.

*COMBINED PISTONS AND RODS*
- To be of Nickel Chrome Steel, heat-treated, to specification.
- The piston and rod to be made in one piece, the forging to be first machined with a surface allowance of ⅛in all over.

**LEFT** Combined piston and piston rod, forged as a single unit in order to reduce reciprocating weight.

Connecting rods could last the life of a locomotive, but in the early 1950s under its regional CME, K.J. Cook, who was ex-Swindon, Doncaster re-designed the outside rods of numerous Eastern Region locomotive classes in order to incorporate GWR-type pressed-in big end brasses. The revised drawing for the standard A3/A4 rod was dated August 1954, and the rods on No. 4472 are to this modified pattern. Being BR renewals they are therefore only inscribed '60103', whereas *Mallard* retains its original rods which are marked '4468'. Cook also instituted highly effective refinements to the vulnerable middle big end bearings, which were accommodated on the existing rods on both classes.

**BELOW** Outside right-hand slidebar stamped '103' and '4472'. The author has personally not been able to verify a suggestion that one of the nine slidebars is an original stamped '1472'.

*WHEELS*

- Wheel Centres to be of Cast Steel to LNER Specification No. 13, Class 'B'. To be perfectly free from honeycomb.
- Coupled wheels to be pressed on axles with a pressure of not less than 95 nor more than 120 tons. Trailing wheels with not less than 80 tons pressure. Bogie wheels with not less than 70 tons pressure.
- The wheel press must be fitted with a reliable automatic pressure recorder, and diagrams must be supplied showing the pressures of all wheels pressed on to axles.
- All wheel centres to have makers' name cast on inside of one spoke.

*TYRES*

- Tyres to be put on after wheels are pressed on axles.
- Tyres to be manufactured from the highest quality of steel by the Acid Open Hearth process, to LNER Specification No. 4, Class 'C' for Coupled Wheel tyres, and Class 'D' for Bogie and Trailing Wheel tyres.
- All tyres to be accurately turned to gauge to section shown on the drawing, and coupled wheels secured to the wheel centres with turned rivets as shown. All tyres to be bored with a shrinkage allowance of 1/800th of the inside diameter.

*AXLES*

- All straight axles to be made to LNER Specification No. 2.
- Built-up Crank Axle. Shafts to be to LNER Specification No. 2.
- Webs, to special specification attached, to be shrunk on the shaft and crank pin, and to be secured by screws as shown on the drawing.

**BELOW** A 4-6-2 driving wheelset, including balanced crank axle, an assembly which weighed 4.8 tons complete with tyres. The leading and trailing coupled wheelsets with plain axles each weighed some 1½ tons less.

# Wheels and tyres

## Wheels

The 4-6-2 driving and coupled wheels, with new tyres (3in thick) shrunk on to the cast steel centres, measured 6ft 8in diameter on tread, which was the Doncaster standard for express passenger locomotives for 50 years from the first 4-4-2 in 1898. Of the major railway works, only Crewe possessed a steel foundry and so for its 4-6-2s Doncaster Works obtained its coupled wheel centre castings from The Darlington Forge Co. Ltd, which also made the necessary wooden patterns, initially for GNR Nos 1470/1.

### BALANCING

The coupled wheel castings incorporated crescent balance weights to counter-balance 1 ton of flailing steel. The latter was comprised of the rotating masses, mainly crank pins, and coupling rods, which in service, could rotate at up to 7 revs/second on a 13in throw, plus the weight of the *outside* reciprocating masses, i.e. pistons, piston rods, crossheads and a portion of the connecting rods, the latter being part rotating and part reciprocating. The shorter middle rod was partially counter-balanced by extensions to the crank webs.

Steam locomotive balancing amounted to a trade-off of riding qualities versus hammer blow on the track, but like a three-legged stool, a three-cylinder locomotive is largely self-balancing, resulting in low hammer blow. The calculations made for the A1 in July 1921 allowed for 60 per cent of the weight of the reciprocating parts to be balanced. The report of the government-sponsored Bridge Stress Committee, published in early 1929, tabulated a modest hammer blow of only 2.6 tons at 6 revs/ second for the A1/A3s (compared with 10 tons by the two-cylinder Ivatt 4-4-2s). Nevertheless, as a result, promptly from April 1929, the proportion on the 4-6-2s was reduced to only 40 per cent.

## Tyres

Tyre fixing was a very precise matter which involved *shrinking* the tyre on to the wheel centre casting. The tyre was uniformly heated by a circular battery of gas burners to bring it to

Heating the new tyre to expand it.

Offering the wheelset to the tyre.

The wheelset is inserted into the expanded tyre.

A view of the entire wheelset as the tyre cools and contracts.

'black heat' in order to expand it sufficiently to allow the insertion of the wheel centre casting (already set on its axle). The shrinkage allowance specified by the LNER was 1/800th of the inside diameter of the tyre, which for a Gresley 4-6-2, amounted to ³⁄₃₂in. The 1923 LNER A1 4-6-2 Specification had also stipulated:

- Wheel centres to be turned to the dimensions on drawing.
- All tyres to be bored smaller by the amount of shrinkage required.
- All the tyres on the coupled wheels are to be tested for shrinkage after they have been put on, and all tyres not found with correct allowance are to be taken off.

Locomotive tyre profiles were a complex issue and as newly fitted they were defined with a precision of ¹⁄₆₄in. Following the recent railway amalgamations, in 1927 the British Engineering Standards Association (BESA) issued new unified standards for British locomotive tyres, which were specified to be 5½in wide, tapered at 1 in 20 on tread, and (normally) provided with flanges 1⅛in deep.

Tyre re-profiling was usually necessary about every 50,000 miles, which would have been covered in merely seven to eight months by the A1/A3s during their pre-war heyday. This operation was undertaken at major running sheds equipped with wheel lathes, thus avoiding any need for visits to Doncaster or Darlington Works. New tyres when fitted,

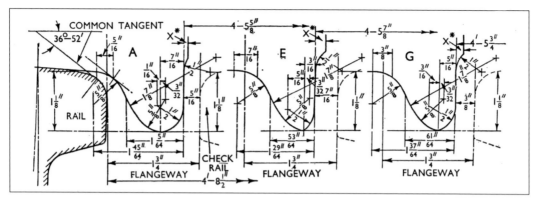

were 3in thick, and considered good for about
400,000 miles. They would be turned down
until a minimum thickness of 1¾in was reached,
when they were scrapped.

**RIGHT** Coupled wheel
axlebox in situ on a
wheelset from No.
4472.

**RIGHT** The aperture
in the frames for one
of the coupled axles,
around the inside of
which are bolted the
hornguides, within
which the axleboxes
engage and are
permitted a measure
of vertical movement
under powerful spring
control.

**RIGHT** A bogie
axlebox engaged in its
hornguide.

# Axleboxes

Irrespective of speed, over the course of a run
from London to Edinburgh the three-coupled
wheelsets of a 4-6-2 would each rotate some
100,000 times (or 1.5 million revolutions for
all ten wheelsets on the engine and tender).
The coupled and driving axle journals (bearing
surfaces) each measured 9½in diameter by 11in
long, and were not enlarged on the heavier A3
class despite a resulting increase in bearing
pressure of nearly 20 per cent. Following the
LNER A1/GWR 'Castle' exchanges in 1925, the
original coupled wheel axleboxes on the A1s
were replaced by those of GWR design (on No.
4472 during its early 1928 general repair).

## SPECIFICATION (1923)

*AXLEBOXES*

- Axleboxes for Bogie and Coupled Wheels
  to be made of Bronze machined all over,
  accurately fitted and recessed in the
  crown.
- The trailing axleboxes to be steel castings
  to LNER. Specification No. 13, Class 'A'
  fitted with a bronze journal bearing, the
  wearing surface to be lined with the
  same white metal as the coupled and
  bogie wheel axleboxes. The top of the
  trailing axlebox to be fitted with slides on
  the 'Cortazzi' principle as shown on the
  drawings.
- Armstrong Oilers to be fitted to all
  axleboxes, and the coupled wheel
  axleboxes to have forced feed
  lubrication.

RIGHT The rear Wakefield mechanical lubricator serving the coupled axleboxes. Also visible is the heavy cover to a sandbox filler.

## Lubrication

Between London and Edinburgh, the three pistons would collectively travel almost 250 miles back and forth within the cylinder walls at an average speed of about 14ft per second at temperatures which were nearly three times hotter than the normal boiling point of water. All this called for extremely effective and reliable lubrication. The 1923 Specification stipulated two six-feed mechanical lubricators (each with 1½–2 gallons capacity) be provided to service the coupled wheel axleboxes and the cylinders.

The A1s were initially fitted with a variety of proprietary lubricators. In 1924, No. 4472 had the Wakefield mechanical lubricator serving the coupled axleboxes, and the Detroit hydrostatic pattern (situated in the cab and mounted on the back of the firebox) distributing high flash point oil to the cylinders. It was provided with the Wakefield type for both purposes in early 1928, an arrangement which had become the standard on all 4-6-2s by 1939, with these mounted together on the right-hand running board.

## Sanding

The antithesis of lubricating oil was sand (fed on to the rails by downpipes) which was utilised on steam locomotives to augment their grip on the rails when they were starting from rest with a heavy train. On the early 4-6-2s, steam-assisted sanding was initially provided in

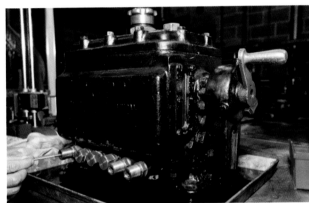

ABOVE The forward Wakefield mechanical lubricator, serving the cylinders, after refurbishment.

front of the leading coupled and centre driving wheels, but this proved very unsatisfactory on the leading set and so was quickly replaced by GWR-pattern gravity sanding. Under British climatic conditions the rails could frequently be damp and slippery, and it was considered that the adhesion factor, i.e. adhesive weight/starting tractive effort ideally should be equal to at least 4. In both the A1 and A3 it was 4.5.

## Suspension

Although British railway track was laid to a high standard, any inevitable irregularities had to be accommodated by allowing a vertical movement of up to 1½in by the axleboxes within the lubricated hornguides bolted to the frames, under the control of powerful springs. In No. 4472 the coupled axlebox springs were of the laminated type with a slight

**SPECIFICATION** (1923)

*SPRINGS*
- Laminated Springs to be to LNER Specification No. 7.
- All Buckles must be made of steel to the quality shown in the Specification, and must be machined out of the solid.
- The plates are to be of Ribbed steel to drawing.

ABOVE Refurbished laminated coupled wheel springs from No. 4472.

RIGHT Driving wheelset (with crank axle). York, January 2011.

FAR RIGHT Driving wheelset below loco frame on the wheeldrop.

BELOW Wheelset being offered up to the locomotive above.

downward camber, having a span of 3ft 6in and composed of 13 plates 5in wide and ⅝in thick held together in a buckle. Laminated springs were mounted externally but above the trailing carrying axle beneath the firebox, and similarly on the tender, while the bogie axleboxes had spiral springs set on each side.

# Brakes

The Gresley 4-6-2s and their tenders were originally equipped with the vacuum brake, the vacuum for both locomotive and train brakes being generated by a graduable steam-activated ejector in the cab. The brakes were held 'off' by the maintenance of a partial vacuum and applied simply by either reducing or destroying this, providing a 'fail-safe' system. The British railway standard working vacuum (the GWR excepted) was 21in of mercury, or 70 per cent of the standard atmospheric pressure of

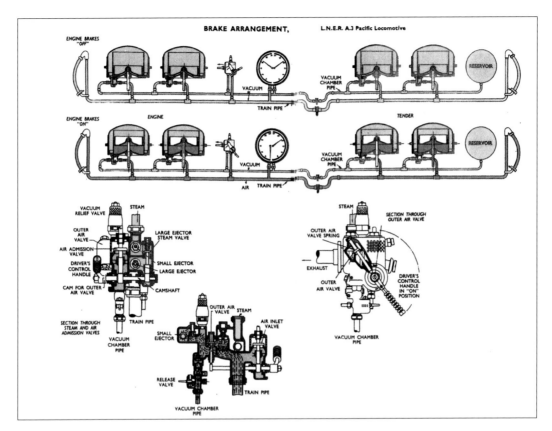

14.7lb/sq in, (which corresponded to 29.2in of mercury). The engine itself was provided with twin 21in brake cylinders, which were mounted between the frames immediately behind and below the middle cylinder casting. These delivered a force of 4.17 tons per lb of atmospheric pressure. Via the compensated brake rigging, a braking maximum force on the engine of about 43 tons (supplemented by the tender) could thereby be brought to bear by renewable cast iron brake blocks acting directly on the steel coupled wheel tyres.

### POST-1963

Remarkably, albeit hauling its own vacuum-braked rolling stock, No. 4472 was not equipped with the air brake for its visit to North America, where the vacuum brake was totally alien and the air brake ruled supreme. In the air brake, which was only universally adopted in Britain for passenger rolling stock in the post-

steam era, the braking force is applied positively by compressed air.

Air brakes were, however, temporarily fitted to the engine for its visit to Australia, but were removed on the engine's later return to the UK. During the 1996–99 heavy overhaul at Southall the vacuum ejector was removed and an air-brake system based on that of the Class 37 Co-Co diesel-electric locomotives installed on the engine and for train braking.

When No. 4472 was acquired by the National Railway Museum five years later, after some debate it was decided to reinstate the vacuum ejector. This would enable the engine to haul steam-era passenger stock on Network Rail and on heritage railways, which particularly tended to employ ex-British Railways 1950s vintage vacuum-braked Mark 1 vehicles. A new air-brake system was also installed to comply with current Network Rail requirements, which is described in Chapter 9.

**ABOVE The original vacuum brake arrangement on the LNER A3 4-6-2 and tender.**

(Original 180lb boiler pressure design)

### BOILER

- To be made strictly to Drawing supplied.
- Boiler barrel, firebox casing, smokebox tubeplate, dome, rivets etc. to be made of Steel made by the Acid Open Hearth process to LNER Specifications Nos 22, 23 and 26.
- Barrel to be made in two plates, telescopic, with flanged tubeplate as shown.
- All joints to be hydraulic riveted where possible with Steel rivets as shown on the Drawings.
- All rivet holes to be drilled to come true to each other, and on no account to be drifted. Sharp edges of holes after drilling to be removed.
- All flanged plates to be annealed after flanging and machining.
- Plates to be all planed up true on edges and carefully fullered on both sides. The surface of plate at joint not to be nicked or damaged in any way by the tool.
- Foundation ring of firebox to be of forged steel, double riveted.

### BOILER TEST

- The boiler before being clothed is to be tested, (with) all fittings, studs, etc., to be in place before testing.

#### Hydraulic Test

- Test pressure to be not less than 235lb per square inch or more than 270lb per square inch.
- Hot water to be used.
- Fullering or caulking may be done at pressures under 100lb per square inch.

#### Steam Test

- Test pressure to be 190lb per square inch.
- The time to raise steam to test pressure to be not less than two hours.
- Full test pressure to be maintained for not less than one hour.
- Fullering and caulking may be done at pressures under 100lb per square inch.

### FIREBOX

- Firebox to be of Copper to LNER Specification No. 16, Class 'A'.

- Tubeplates 1in thick at tube part, other part ⁹⁄₁₆in thick.
- Copper stays to be in accordance with LNER Specification No. 17 and to be tightly screwed into plates and sawn off to length; to be hammered over with pneumatic hammer and three-pronged sett with centre.
- Plates to be planed up on both edges, and joints carefully caulked inside and out.
- Particular attention must be given to the workmanship, especially for bottom corners of outer casing and firebox.
- The roof of box to be stayed to outside shell with stay bolts screwed in outside shell and riveted over, and screwed tightly into copper plates with nuts inside, as shown on drawing.
- Transverse and longitudinal stays to be fitted as shown on drawing.
- Flanging blocks for steel and copper plates for firebox and smokebox tubeplate will be supplied by the Railway Company.

### SMOKEBOX

- Plates to be Mild Steel, to LNER Specification No. 25.
- Door to be flanged, and to be a perfect joint on smokebox.

### SUPERHEATER HEADER

- Superheater Header to be of Cast Iron to LNER Specification No. 14, the same quality as for Cylinders, to be perfectly free from honeycomb and sound throughout, and tested for leakages by hydraulic pressure of 250lb per square inch before fitting into smokebox.

### TUBES

- Superheater and Boiler tubes and Superheater Elements to be cold drawn seamless steel tubes to LNER specification No. 20.
- Tubes to be secured to tubeplate in the manner shown on the drawings supplied.

### ASHPAN

- To be made of Steel Plates of the same quality as the smokebox.

# Boiler and firebox

## Boiler

The GNR 4-6-2 boiler in 1922 was the largest yet built for a British locomotive, particularly as regards its maximum outside diameter at the firebox end of 6ft 5in, which was exceeded only by ½in in those of the last British Pacifics built in 1954.

The original GNR 4-6-2 boiler was designed to work at the moderate pressure of 180lb per sq in, and a pair of compact Ross 'pop' safety valves were set to 'lift' when this pressure was exceeded. British locomotive boilers in general, however, were designed to incorporate a safety factor of about 5, so that the specified plate thicknesses and riveted joints etc. of the 4-6-2 boiler would have been designed to withstand a pressure of at least 900lb. This made due allowance for an acceptable measure of inevitable wasting of the barrel and firebox plates before replacement of the boiler eventually became due.

## Firebox

When compared with the relative simplicity of the cylindrical boiler barrel, the firebox was a highly complex affair. The inner firebox was constructed from very low tensile (particularly

at high temperatures) but highly conductive copper, which was secured at a distance from the steel outer casing, with an intervening water space, by nearly 2,000 threaded copper stays. It was subjected at normal working pressure to a combined potential *imploding* force of almost 2,500 tons. Not only that, but the ⁹⁄₁₆in thick copper plates also experienced on the water side, a temperature

**LEFT** A new A3 4-6-2 boiler photographed at Doncaster Works in November 1929. The steel crinolines held the asbestos insulation strips in place and provided a foundation for the steel boiler clothing plates.

approaching 400°F (the approximate elevated
boiling point of water at usual locomotive
boiler working pressures). While on the other
side, temperatures at virtually its own melting
point (1,980°F), i.e. of the order of 2,000+°F
(yellow heat), could prevail within the firebox
interior when the locomotive was working
hard. In order to minimise the still invariably
catastrophic consequences if its crown
became uncovered, resulting in the collapse of
the inner firebox, a fusible plug of low melting-
point alloy was set in at its highest point.

## Boiler construction

The hundreds of rivet holes were centre
punched and pre-drilled in the barrel and firebox
plates when flat. The barrel was created by
repeatedly passing the plates cold through a
grouping of three or four steel rollers, which
were pre-set with respect to each other in order
to allow for the plate thickness and to evolve
the barrel diameter(s) required.

The rear course of the 4-6-2 boiler consisted
of a roughly trapezoidal shaped steel plate
which was ¾in thick and measured 19¾ft wide
(maximum) by nearly 8¾ft long, whose passage
between the rollers was also required to impart
a 1-in-15 taper to the resulting barrel. By coning
the barrel down from 6ft 5in to 5ft 9in outside
diameter, weight was concentrated where it was
most effectively utilised, i.e. at the 'business'
or firebox end. This also improved the severely
restricted view ahead from the cab.

The outer firebox back and throat plates
on the other hand, were shaped when red
hot on pre-formed iron castings or flanging
blocks mounted in a hydraulic press. The barrel
courses were temporarily tacked together with
bolts, and the entire boiler was suspended
vertically while the matching holes in the
overlapping plates were reamered (smoothed
out) prior to the red-hot rivets being inserted.
The riveting process was an ear-splitting
operation, which was undertaken without the
benefit of any ear protection. (In the recent
boiler for *Tornado*, rivets were largely eliminated
by resorting to extensive welding.)

The copper inner firebox, with forged steel
foundation ring already attached, would then be
inserted through the bottom of the steel outer
firebox, their respective stay holes having been
pre-drilled undersize. These had to correspond
exactly with each other prior to their being
screw-threaded with taps and the insertion
of the threaded stays themselves, having
protective nuts applied on the inside, and then
beaded over on the outside.

## Insulation

On completion, the boiler was then insulated
or 'lagged' in order to reduce significant heat
loss, as its temperature when in operation
would be about 350°F above mean ambient.
On a boiler of this size the yearly fuel saving

thereby effected was of the order of 50 tons of coal. In earlier times, lagging simply consisted of strips of timber, but by the 1920s and '30s the favoured locomotive boiler insulator was asbestos (which, in 1923 for the LNER 4-6-2s was specified to be of the Bell's 'Wadnit' variety). At that time the very serious health-threatening properties of this then ubiquitous material were not acknowledged. In later years, glass fibre, itself also distinctly unpleasant to handle, was sometimes substituted for this purpose. All was contained within a sheet steel casing.

## Superheater

Although at normal firing rates of 50–60lb per square foot of grate area, a locomotive boiler could convert say 75 per cent of the heat energy in the coal into steam, only 5 to 7 per cent of this would actually emerge at the tender drawbar to pull the train. This was mainly because around 70 per cent of the total heat energy in steam is 'unusable' latent heat of evaporation, and so the cylinders at best can only realistically convert no more than about 15 per cent into mechanical work, a portion of which is further absorbed in propelling the locomotive alone.

However, even doubling the cylinder efficiency from say, 6 to 12 per cent, will halve the coal bill. By superheating the steam, i.e. raising its temperature from saturation at $c$400°F (due to the increased pressure) to say 650°F, in addition to expanding its volume by 40 per cent, injects more 'sensible' or usable heat into the steam, which significantly increases cylinder efficiency and therefore power output, *and* reduce coal consumption.

Superheating was adopted very rapidly in Britain after 1910, and the Gresley 4-6-2s were equipped with the Robinson-type superheater. Saturated steam passed through 1¼in bore steel elements which were quadrupled in the rear half of the flue tubes closest to the firebox. By 1920, it was no longer considered necessary to fit pyrometers and damper mechanisms with which the driver could monitor, and if necessary restrain, steam temperatures from rising *too* high (say above 650°F), to prevent perceived serious cylinder and valve lubrication problems. As it proved, steam temperatures did not rise high enough on the Gresley A1 and A3 4-6-2s!

A small circular anti-vacuum or 'snifting' valve, mounted on the top of the smokebox immediately behind the chimney, when coasting with steam shut off, prevented smokebox gases and ash from being sucked into the cylinders. It also permitted air to circulate through the superheater elements to prevent them from burning.

**ABOVE** A new 4-6-2 boiler lagged with asbestos at Doncaster Works in 1932.

## Draughting

The actual dimensions and relative proportions of the blastpipe orifice and chimney liner (or petticoat pipe) are crucial, but can be altered relatively easily in order to improve steaming. The discharge of the exhaust steam from the cylinders through the blastpipe creates a very slight vacuum within the smokebox, the suction being equivalent to only (minus) a few inches of water (whereas atmospheric pressure

**BELOW** No. 4472's superheater header (with cover plates removed) and elements in place.

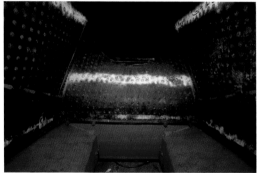

ABOVE A close-up of the forward portion of firegrate, showing transverse firebars of the 'drop' section immediately behind the throat plate. (This view was taken inside the very similar A4 firebox.)

ABOVE RIGHT Inside the new A3 firebox, the new ashpan is seen in place below the foundation ring, before the later insertion of the firebars. The extended 'pegs' protruding from the sides of the inner firebox will carry the brick arch, which is essential for the combustion process within the firebox.

BELOW Arrangement drawings of LNER 'banjo' dome.

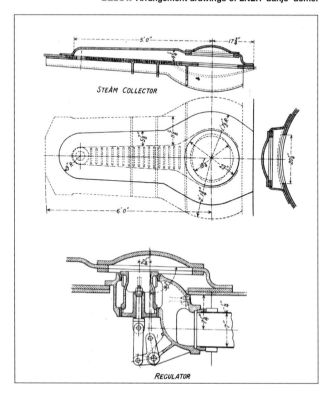

STEAM COLLECTOR

REGULATOR

equates to *plus* 33 *feet* of water). 1lb of exhaust steam evacuates 1½ to 2lb of hot gases from the smokebox and thereby draws on the fire. The complete combustion of 1lb of coal requires 15lb, or about 200cu ft of ambient air, whose supply to the grate can also be regulated by the fireman via his manipulation of the hinged horizontal damper doors in front of the ashpan.

## Ashpan

The importance of the humble ashpan could not be understated. Fortunately, the ashpan designed for the GNR 4-6-2s in mid-1921 proved to be adequate to accommodate the ash residue which would later accumulate during the course of exceptionally long through runs of almost 400 miles, but which could not at that time have been anticipated.

Ash falls into the ashpan from the incandescent firebed, which is supported by the replaceable cast iron firebars, through the air spaces provided between them, which in the Gresley 4-6-2s amounted to 37 per cent of the grate area.

Another more troublesome residue from the combustion, vitrified clinker, would remain lying on the grate. In order to assist with its disposal at the end of the working day, the front firebars on the Gresley 4-6-2s were arranged transversely and as a hinged drop grate. From this, the clinker, after being pushed forward by long fire irons pushed through the firehole, could then be tipped down into the ashpan from which it would then be laboriously removed, with the ash.

## Boiler development

The basic A1 4-6-2 boiler assembly (without the smokebox) measured 28½ft long overall, increasing in length by about ¾in when in

| COMPARATIVE LEADING DIMENSIONS OF GRESLEY 4-6-2 BOILERS | | | |
|---|---|---|---|
| Diagram No. | 94 | 94HP/94A | 107 |
| Class fitted, introduced | A1, 1922 | A1, 1927<br>A3, 1928/1934 | A4, 1935<br>A3, 1954 |
| Built during | 1921–26 | 94HP, 1927–35/46–47 | 1935-61 |
| | | 94A, 1934–46/49–50 | |
| Working pressure, lb/sq in | 180lb | 220lb | 250lb (220lb on A3) |
| Maximum diameter | 6ft 5in | 6ft 5in | 6ft 5in |
| Length between tubeplates | 19ft 0in | 18ft 11¾in | 17ft 11¾in |
| Firebox outside length | 9ft 5½in | 9ft 5¾in | 10ft 5¾in |
| No. @ dia. flue tubes | 32 @ 5¼in | 43 @ 5¼in | 43 @ 5¼in |
| No. @ dia. fire tubes | 168 @ 2¼in | 125* @ 2¼in | 121 @ 2¼in |
| Heating surface (sq ft): | | | |
| Tubes & flues | 2715 | 2522 | 2345 |
| Firebox | 215 | 215 | 232 |
| Total evaporative | 2930 | 2737 | 2577 |
| Superheater | 525 | 706 | 749 |
| Grate area (sq ft) | 41.25 | 41.25 | 41.25 |
| Empty weight (tons) | 23.9 | 26.6/27.0 | 28.0 |
| Fitted to No. 4472: | 1923–46 | (94HP) 1947–48<br>(94A) 1948–77, 2015– | 1978–2005 |

\* Later reduced to 121

steam due to the thermal expansion of the steel plates, which had to be allowed for. Weighing 26 tons empty, the subsequent higher pressure (220lb) version was very similar but, due to slightly thicker plates and its larger superheater, it weighed 2.4 tons.

The original 180lb boilers were designated Diagram 94, while the initial 220lb versions, likewise fitted with conventional circular domes, were Diagram 94HP. Diagram 94A covered 220lb boilers provided with the distinctive elongated so-called 'banjo' dome, which made its 'invisible' debut in May 1934 on 2-8-2 No. 2001 Cock o' the North.

No. 4472 was one of the last A1s to be converted to A3, in January 1947. The last was BR No. 60068 Sir Visto in November 1948, which resulted in a stock of 78 A3s, the pioneer Great Northern having been very extensively, and controversially, rebuilt by Edward Thompson in 1945 (when the remaining Gresley A1s were re-classified A10).

No. 4472 was routinely fitted with a total of 15 different boilers at Doncaster Works between

LEFT **Plan view of regulator valve on No. 4472 housed in 'banjo' dome.**

**BOILERS FITTED TO LNER NO. 4472, 1922–62** (all fitted at Doncaster Works)

| Date | Boiler No. | Type | First use | Previously fitted to 4-6-2: |
|---|---|---|---|---|
| February 1923 | 7693 | 94 | 2/1923 | New boiler (built 1922) |
| April 1928 | 7878 | 94 | 4/1928 | New boiler (built 1926) |
| April 1933 | 7804 | 94 | 7/1924 | LNER No. 2581 Neil Gow |
| May 1935 | 7772 | 94 | 12/1924 | LNER No. 4471 Sir Frederick Banbury |
| November 1939 | 7785 | 94 | 7/1924 | LNER No. 2561 Minoru |
| January 1947 | 8078 | 94HP | 10/1928 | LNER No. 2576 The White Knight |
| March 1948 | 9119 | 94A | 5/1941 | Ex-LNER No. 40 (1946) Cameronian |
| December 1949 | 9448 | 94A | 4/1944 | Ex-LNER No. 93 (1946) Coronach |
| March 1952 | 27015* | 94A | 6/1941 | BR No. 60047 Donovan |
| April 1954 | 27074* | 94A | 7/1946 | BR No. 60082 Neil Gow |
| October 1955 | 27007 | 94A | 10/1950 | BR No. 60077 The White Knight |
| July 1957 | 27011 | 94A | 11/1950 | BR No. 60054 Prince of Wales |
| January 1959 | 27044* | 94A | 6/1949 | BR No. 60097 Humorist |
| August 1960 | 27047* | 94A | 8/1944 | BR No. 60100 Spearmint |
| June 1962 | 27058* | 94A | 12/1939 | BR No. 60037 Hyperion |

*numbers after 1950 boiler re-numbering

**HEAVY REPAIRS CARRIED OUT TO DIAGRAM 94A BOILER NO. 27021**
(Carried by eight A3s between 1949 and 1964)

| Date | New flues @ 5¼in (steel) | New tubes @ 2¼in (steel) | New firebox stays fitted (copper) | Notes |
|---|---|---|---|---|
| May 1949 | (43) | (121) | (1810) | New boiler, into service July 1949 on Colombo |
| December 1950 | | | 560 | |
| March 1952 | | 121 | 926 | |
| August 1953 | 43 | 121 | 293 | |
| August 1955 | | 121 | 1712 | New copper firebox & outer casing, also new smokebox tube plate |
| August 1957 | 22 | 121 | 808 | |
| February 1959 | | | 740 | |
| September 1960 | | | 584 | |
| November 1962 | 36 | 121 | 890 | |
| November 1964 | | | | Boiler on Lemberg, when loco withdrawn |

**HISTORY OF DIAGRAM 94A BOILER NO. 27020**
(Renumbered from 9454 in 1950, boiler data is courtesy of Melvin Haigh)

| Period | Class A3 4-6-2 locomotive | Recorded mileage |
|---|---|---|
| Dec 1944–Dec 1948 | New boiler and firebox (No. 1) on LNER No. 2570 Tranquil | 222,790 |
| Feb 1949 | fitted with new firebox (No. 2) | |
| Mar 1949–June 1952 | on BR No. 60073 St Gatien | 139,662 |
| July 1952–Apr 1954 | on BR No. 60069 Sceptre | 90,598 |
| June 1954–Feb 1956 | on BR No. 60097 Humorist | 106,036 |
| Feb 1956–Mar 1957 | stored at Doncaster Works, fitted with new firebox (No. 3) | |
| Mar 1957–Mar 1959 | on BR No. 60070 Gladiateur | 81,909 |
| May 1959–Feb 1961 | on BR No. 60110 Robert the Devil | 92,991 |
| Mar 1961–Dec 1962 | on BR No. 60041 Salmon Trout | not recorded |
| Dec 1962–Mar 1965 | stored, later sent to Darlington Works | |
| Mar 1965–Dec 1977 | on (LNER) No. 4472 Flying Scotsman | c200,000 (est.) |
| 1978–2009 | stored and later repaired | |
| 2010 | fitted with new firebox (No. 4) at Bury | |
| 2015– | on (BR) No. 60103 Flying Scotsman | |

1923 and 1962, which had been built between 1922 and 1950, although of these only the first two to be fitted, both of the original 180lb type, had been new.

Once built, a boiler would still subsequently require periodic heavy maintenance work, from the routine replacement of the fire tubes, flues and firebox stays, to partial or even total renewal of the firebox. (Superheater elements would sometimes be replaced on shed.) This was undertaken in the boiler shop, usually after the boiler had been removed from one locomotive, and before it was put back on the frames of another. The heavy repair history is given on the facing page for boiler No. 27021, identical with that now carried by No. 4472 (No. 27020), for which the comparable record regrettably has not survived. Despite being consecutively re-numbered in 1950, No. 27021 was actually built five years later.

Surviving records show that several A3 and A4 boilers typically had their copper fireboxes replaced at four- to five-year intervals and that some were successively fitted with as many as four replacement copper fireboxes, and one or two new outer casings. The main boiler barrels sometimes enjoyed an overall working life of nearly 30 years, in some instances lasting from the mid-1930s until the early 1960s, by which time their annual mileage had very significantly reduced.

**POST-1963**

When restored at Doncaster Works in early 1963, No. 4472 initially simply retained the rather elderly 1939 vintage Diagram 94A boiler which had been fitted during its last heavy repair there, only eight months previously. In November 1963, Doncaster ceased to carry out steam locomotive repairs, and one year later in November 1964, No. 4472 entered Darlington Works where it underwent a boiler change. It emerged in March 1965 with a slightly newer, 1944-built boiler (No. 27020), which had last seen service on and had been removed from No. 60041 *Salmon Trout* at Doncaster in December 1962.

This boiler was then carried until the 1978 Vickers re-fit, when a Diagram 107 boiler (No. 27971), which had been built at Doncaster Works as late as 1960, was substituted. This had been fitted new to A4 No. 60017

## LOCOMOTIVE BOILERS – INSTRUCTIONS FOR TESTING, EXAMINATION & REPAIR

(Revised, June, 1949)

### Periodical Examination

■ Each locomotive boiler is to be thoroughly examined at least once every six months by the Mechanical Engineer's Inspectors.

■ All wash out plugs, hand hole and mud hole doors and fusible plugs to be taken out and thorough examination for sediment and scale in water spaces to be made.

■ Firebox stays, roof stays and bolts to be hammer-tested and thorough examination of firebox made. The brick arch must be taken down for this to be done.

■ The smokebox tubeplate to be thoroughly cleaned and examined.

■ Water gauges and water ways to be examined.

■ Pressure gauge to be tested against a Master gauge.

### Complete Examination

■ Boilers must be completely examined every five years or more often, at the discretion of the Mechanical Engineer.

■ All clothing must be removed, a thorough internal and external examination carried out and a special record kept of such examination. Whenever possible, complete boiler examination should be made in the Works whilst the locomotive is undergoing general overhaul.

### Safety Valves

■ No adjustment to be made to 'Ross' safety valves by Motive Power Department staff. If an alteration is required, application must be made to the Mechanical Engineer.

■ A spare valve to be substituted and the defective valve to be returned to the Works for repairs.

**A.H. Peppercorn**
Chief Mechanical Engineer
The Railway Executive
Eastern & North Eastern Regions
Doncaster

*Silver Fox* at Doncaster in October 1960, and subsequently in April 1962 to sister engine No. 60019 *Bittern* (itself also now preserved), from which it had been removed at Darlington in February 1965. After that, it had been purchased by Alan Pegler for possible future use on No. 4472. In 2005, the National Railway Museum decided to reinstate A3-type boiler No. 27020, whose history is given opposite.

# Cab and controls

The Gresley 4-6-2 cab would have appeared positively luxurious to men who had hitherto been accustomed to the vestigial shelters of the Ivatt 4-4-2s, which, with their two cylinders, very short rigid wheelbase and heavy back ends, were notoriously rough riding. Unusually, the 4-6-2s were designed to be driven with the driver sitting down, with a slightly smaller perch even being provided for the fireman as well. This humane approach can be traced to Gresley's brief spell in the Horwich drawing office of the Lancashire & Yorkshire Railway *c*1898, when the distinctive Aspinall 'High Flyer' inside-cylinder 4-4-2s were being designed with a similar unusual degree of enlightenment for that period.

Also, compared with the earlier 4-4-2s, the 4-6-2s with their longer rigid wheelbase, three-cylinder propulsion, and Cortazzi slides for the carrying axle beneath the firebox, all made for much smoother riding characteristics. This would have made life a little easier for the fireman, who was otherwise confronted with stoking a firegrate which was some 35 per cent larger than anything with which he had previously been accustomed (41¼sq ft *v* 31sq ft as on the 4-4-2s). The 4-6-2s had otherwise not been entirely welcome when first introduced in their original short-valve travel coal-eating form.

Vertical regulator handles were provided on each side of the cab (another feature of the initial batch of LYR 4-4-2s), to facilitate backing on to a train. Analogous to the accelerator pedal on a motor vehicle, these were directly connected to the regulator valve itself, or 'throttle', in the dome, which controlled the flow of steam from the boiler through the superheater to the cylinders. The reverser, linked to the valve gear, was originally located on the right-hand side of the cab, as both the Great Northern and North Eastern railways favoured right-hand drive, despite the fact that a left-hand (i.e. 'near-side') driving position made for better sighting of signals etc. In earlier years, when boilers were of smaller diameter and lower pitched, this had been less of a problem.

All 52 Gresley A1 4-6-2s had originally been built with right-hand drive. Despite representations from footplate staff, particularly

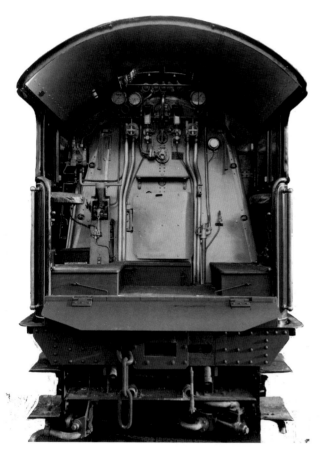

**ABOVE** A cab view of a new LNER A3 4-6-2 in 1928 (with left-hand drive). A hinged sheet-metal screen obscures the firehole door. The steam soot blower in the centre of the back plate was later removed from the 4-6-2s in the mid-1930s. More comfortable bucket seats for both crew were later substituted, and this *general* layout applied to No. 4472 only from 1954. Around 1960 an AWS indicator and a speedometer dial also made their appearance.

## SPECIFICATION (1923)

*CAB*

- Plates to be Mild Steel, to LNER Specification No. 25, with rivets countersunk on outside.
- Hinged windows to be fitted to front plate, and sliding windows in the sides.
- Seats and wood platform to be provided as shown on drawing.

**RIGHT A 2005 view of the driver's side of the cab.**

in Scotland, in the early 1930s, and strong recommendations from the Ministry of Transport, surprisingly their actual conversion to left-hand drive to conform to the new-built A3s, was not formally approved until nearly 20 years later, by British Railways as late as 1951. The alteration was actually carried out between September 1952 and July 1954 (on No. 4472 in April 1954). A curious permanent legacy of the early right-hand drive, which furthermore was perpetuated on all subsequent LNER Pacifics (even including *Tornado*), was that the lever controlling the cylinder drain cocks still remained on the right-hand side of the cab. Paradoxically, that for the sanding gear *was* moved across to the driver's new position.

The 4-6-2s were fitted with reversers of the vertical 'capstan' type, which had featured on the second (1902) series of LYR 4-4-2s, which permitted fine adjustments to the cut-off in the cylinders. Cut-off was indicated on a graduated (but not necessarily very accurate) vertical brass indicator strip mounted on the back of the firebox. With a good head of steam when starting away from rest with a heavy train, the reverser would be set at or close to full forward ('bottom') gear. To open the regulator valve too sharply could lead to severe slipping and 'tearing' the fire, but the regulator would then be steadily opened while the cut-off was progressively reduced as speed increased.

Ideally, a locomotive would be worked on full regulator to avoid throttling the steam, with a minimum cut-off of perhaps 15 per cent, although 25 per cent was typical. The actual combination of cut-off and regulator opening depended on the line speed, trailing load and the prevailing gradient, rising or falling (when the regulator might be closed). Steam locomotive driving techniques varied widely and were expertly discussed by G.A. Weeden in *Trains Illustrated* magazine for March, April and May 1960.

The adoption of a form of continuously variable reverser was a considerable advance over the clumsy lever mechanism having only stepped notches on the 4-4-2s, which was

**ABOVE The fireman's side of the cab in 2005, showing the duplicate regulator handle provided for when running in reverse.**

**LEFT Graduated brass cut-off indicator.**
*(Author)*

RIGHT A gauge glass contained within its armoured glass 'protector'.

FAR RIGHT Boiler steam pressure gauge.

difficult to manipulate at speed and therefore prompted drivers to leave them set at a fairly late cut-off. Consequently, they preferred to drive 'on the regulator' and in the absence of 'expansive working' by not shortening the cut-off this resulted in increased fuel consumption, and thereby an unnecessary additional burden for their firemen. Power-assisted reversers, although particularly popular on the former North Eastern Railway and in Scotland, were not favoured at Doncaster.

RIGHT Live steam injector.

BELOW Exhaust steam injector.

The cab also contained controls for the boiler water feed from the injectors, ashpan dampers to regulate the primary air supply to the firebed, sanding and vacuum ejector brake valves, and for train steam heating. In addition, there was simple monitoring equipment, i.e. twin gauge glasses which directly (and crucially) displayed the water level in the boiler, and dial gauges showing the steam pressure in the boiler and steam chests, and the brake vacuum.

Using steam jets and having the virtue of no moving parts, the injectors ingeniously forced feed water into the boiler against its own prevailing internal pressure. These were located beneath the cab floor, on each side of the engine between the inner and outer frames, with the Davies & Metcalfe No. 10 exhaust steam injector on the right-hand side, and the Gresham & Craven No. 11 combination live steam injector on the left. These relative positions had been transposed when No. 4472 had been converted from right- to left-hand drive to ensure that the *exhaust* steam injector remained on the fireman's side of the cab. This obviously could only be employed when the engine was on the move and with the regulator open. It normally recycled about 5 per cent of the exhaust steam whose (largely latent) heat would otherwise have been discharged to waste, thereby pre-heating the feed water and thus resulting in a useful fuel economy.

The *live* steam injector was used to top up the boiler when the engine was stationary and when the regulator was closed. It also supplemented the exhaust steam injector at times of high demand on the A3s, these having a maximum feed rate of 2,500 gallons (25,000lb) per hour. The live steam injector also provided a back-up should the exhaust steam injector fail, as injectors could be notoriously temperamental. Perhaps surprisingly, exhaust steam injectors were by no means universally popular. Feed water heaters, although commonplace abroad, of which the French ACFI type was experimentally fitted by the LNER to a representative A1 and A3 locomotive, were not successful in Britain as they were considered to cost more to maintain than the value of the coal that they saved.

## Safety equipment

In October 1928, No. 4472 paid a unique visit to Gateshead Works to be fitted with Raven cab fog signalling apparatus. This mechanical system was discontinued just five years later and removed. However, the tragic double collision in fog on the West Coast Main Line at Harrow & Wealdstone in October 1952, which resulted in 112 fatalities, prompted the development by British Railways of its Automatic Warning System, or AWS. The necessary fixed equipment was installed between the rails in the 'four foot' on the East Coast Main Line throughout, between London and Edinburgh during 1959–61, and many A3s were fitted up, including No. 4472, at a cost of approximately £300 per engine (£5,400 at 2013 prices).

The AWS was linked to the locomotive and train vacuum brake systems, and provided both an audible and a visual warning on the footplate which was electrically activated on approach when about 200 yards from a yellow distant signal, if it was set at caution. After a slight delay, if the driver failed to take action, a brake application was automatically effected by reducing the vacuum which would halt the train before it passed the ensuing red home or stop signal set at danger about half a mile further on.

Also for safety reasons, somewhat belatedly in early 1960, BR announced that during the next few months it would be fitting 1,007 steam locomotives, including 75 A3s, with speed indicators. Manufactured by Smith-Stone Ltd, as fitted to No. 4472, this took its drive off the trailing left-hand crank pin. The actual dates of fitting of the AWS and the speedometer to No. 4472 at Doncaster Works were not officially recorded. In the post-steam era, both fittings have since become mandatory equipment on

steam locomotives passed for running on the British national railway network.

The lowering of the cab roof on Nos 4470–80 for working in Scotland significantly reduced the size of the cab front windows, and thus the forward visibility from the footplate, already restricted by the unusually large diameter of the boiler at the firebox end. In the process, the profile of the cab side sheets, including the rear cut-out, was also slightly modified. Inevitably, drivers leaned out of the rear side window, for which full height hinged glass sight screens, 4in wide, were provided on the 4-6-2s from 1929.

LEFT AWS/TPWS indicator on No. 4472, mounted on the firebox back plate close to the driver.

BELOW Flexible drive for Smith-Stone Ltd speedometer, originally fitted to No. 4472 c1960.

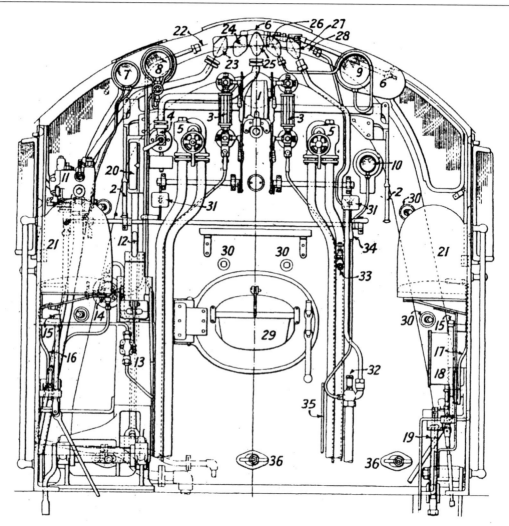

**Explanatory chart of cab fittings of LNER A3 and A4 4-6-2 (left-hand drive).**

1   Regulator stuffing-box.
2   Regulator handles.
3   Water gauges.
4   Blower valve.
5   Combined injector steam
    and feed water delivery
    valves.
6   Stop valve for steam
    stand.
7   Duplex vacuum gauge.
8   Steam chest pressure
    gauge.

9   Boiler pressure gauge.
10  Carriage heating pressure
    gauge.
11  Vacuum ejector.
12  Reversing screw handle.
13  Reversing gear clutch.
14  Steam stand valve.
15  Water control for injectors.
16  Sand-gear lever.
17  Cylinder cock lever.
18  Speed recorder (A4 only
    and removed after 1939).

19  Drop-grate screw.
20  Cut-off indicator.
21  Enginemen's seats
    (driver L H side, fireman
    R H side).
22  Steam sand supply valve.
23  Ejector steam stop valve.
24  (Blank)
25  Blower stop valve.
26  Pressure gauge stop valve.
27  Carriage heating stop
    valve.

28  Mechanical lubricator
    warming valve.
29  Firehole door.
30  Washout plugs.
31  Remote control for water
    gauge cocks.
32  Carriage heating safety
    valve.
33  Coal watering cock.
34  Whistle control.
35  Damper rod.
36  Handholes.

restored in 1963, it disappeared about 20 years later when the tender tank was completely renewed in compliance with the original drawings. The now superfluous water pick-up equipment was then also discarded. Since 1999 this tender has been air braked.

## No. 4472's auxiliary tender, 1966–74

As described later, in 1966 Alan Pegler purchased a second tender (No. 5332) for conversion purely as a water carrier to augment No. 4472's supplies on long runs. Latterly attached to A4 No. 60009 *Union of South Africa*, this was also one of the original 1928 corridor tenders. The corridor was retained but the water capacity was increased from 5,000 to 6,000 gallons at the expense of the coal space, and this remained with No. 4472 until 1974. Thirty years later it was sold to the owner of preserved A4 No. 60019 *Bittern*. On that engine's return to active service in 2007 (with another corridor tender) it was additionally attached so as to increase the locomotive's potential operating range to about 250 miles. Later in 2011, this tender was modified by Roland Kennington who relocated the corridor down the centre in order to improve its stability at speed.

While No. 4472's tender tank was being replaced c1983, corridor tender No. 5324 from A4 No. 4498 *Sir Nigel Gresley* (then also undergoing extended repairs at Carnforth), was temporarily attached to the A3, and was painted *plain* apple green while retaining the stainless steel LNER insignia. This tender had previously been attached to No. 4472 between 1929 and 1936.

## Water treatment

For locomotive purposes, hard water was treated or 'softened' to remove a high proportion of any dissolved calcium and magnesium salts, found particularly in limestone areas. Otherwise this could lead to scaling of heating surfaces, thereby greatly reducing the efficiency of the boiler, and also cause corrosion which could shorten the life of the tubes and firebox, or indeed the entire boiler. However, it was still necessary, when on shed, to 'blow down' the boiler under its own residual steam pressure and to regularly drain it completely. It was then thoroughly washed out with water

**LEFT** A washout door on the upper side of the firebox, set just above the level of the inner firebox crown. On shed, on the removal of the screw threaded plug within, high-pressure water jets would be admitted to scour the crown of any accumulating scale.

**LEFT** The manhole on the underside of the boiler barrel.

jets through ports in the firebox sides and back plate, which were normally sealed by screw plugs or washout doors. Typically, this unpleasant procedure was carried out once a week, taking 10 to 18 hours, often on a Sunday in the case of express passenger locomotives.

The frequency of this irksome task would be greatly reduced by the late 1930s, to perhaps only once a month, by the adoption of continuous or automatic blow-down, to counter the otherwise rapidly increasing concentration of dissolved matter in the boiler. (Between London and Edinburgh a 4-6-2's boiler would effectively be recharged at least five times.) This involved continuously bleeding off a small quantity of water from the boiler, perhaps 1–1½ gallons per mile when in operation, and discharging this into the ashpan or directly on to the track.

No. 4472 was equipped with continuous blow-down in November 1939, only three weeks after having completed a seven-week general repair. On the 4-6-2s, a circular manhole was provided on the underside of the boiler barrel at its lowest point immediately ahead of the firebox, for the removal of sludge. This was a Doncaster feature dating back to Patrick Stirling's time, and can be seen on his 4-2-2, GNR No. 1.

**RIGHT** The locomotive coaling plant at King's Cross depot, pictured in March 1935, about four years after its construction. It would have charged No. 4472's tender on numerous occasions, but was demolished soon after the closure of the depot in June 1963.

## Coaling

Another feature of the steam era, which could be found at the larger locomotive depots, was the automatic overhead coaling plant, which eliminated one aspect of the particularly hard physical work associated with steam locomotive operation (but never from the many small sheds prior to their eventual demise).

A loaded locomotive coal wagon was hauled up the side, and its contents tipped into a large internal bunker, from which prescribed amounts of coal would be discharged into the tender of a locomotive waiting beneath. It took six minutes to charge a 4-6-2 tender, usually when the engine came back on shed. Such a coaling plant was installed at King's Cross shed in 1931 during No. 4472's first sojourn there. A concrete locomotive coaling plant survives at Carnforth, Lancashire.

## TURNTABLES

Another major item of infrastructure required by steam locomotives with tenders, simply to turn them around, was a turntable, usually located at a locomotive shed, or sometimes at stations. A 70ft diameter table capable of accommodating the new 4-6-2s and their tenders was installed just beyond the platform ends in the locomotive yard at King's Cross station in 1924, a facility that was not provided at King's Cross 'Top Shed' nearby, until ten years later. Very few turntables now survive, but one recovered from Gateshead and installed at Scarborough c1980 regularly turned No. 4472 during 2004–05.

**ABOVE AND RIGHT** No. 4472 on the turntable at Scarborough, 27 May 2005.

# Painting, lining and lettering

## SPECIFICATION (1923)

*PAINTING*

- All Ironwork to be thoroughly cleaned, free from rust and grease.
- The Boiler to have one coat of Best Oxide Paint before the lagging is put on, and the lagging plates to be well painted on the inside before being put on.
- The whole of the Engine and Tender to be painted with one coat of Lead colour, 'stop' all bad places and rivets, and then plaster all over with Ferguson's Enamel Filling, and afterwards rub down with Pumice blocks to a level surface. Then to have two more coats of Lead colour and two coats of London & North Eastern Railway Standard Green, picked out, lined and lettered to pattern. Then to have three coats of Varnish, flatted between each coat.
- Frames on outside and Buffers to be painted Black, and lined Red to pattern.
- Buffer Plates to be Red, picked out and lined in White.
- Inside of Frames, Axles, Injectors, Ejectors etc. to be painted with one coat of Spirit Red and one coat of Varnish Red.
- Engine Wheels to be painted Green and picked out Black, and Varnished three coats before being put under engine.
- Brake gear, Spring gear, Ashpan, Footplate, Tender wheels, Scoop gear etc. to have two coats of Japan Black.
- Inside of cab, front of Tender, to be finished Dark Green and Varnished three coats. Top of Cab outside to be Japan Black.
- Smokebox and Chimney to have one coat Asbestos Black, and one coat Japan Black.
- Tender Top and Coal Bunker to be Japan Black.
- Number of Engine to be on front buffer plate and side of Tender, to be lettered to Drawing supplied.

*Additional notes:*

- Pre-1928, LNER in gold, shaded in red, 7½in high, numerals 12in gilt shaded in red.
- Post-1928, LNER in gold, shaded in red, 12in high, numerals 12in gilt shaded in red.

**ABOVE** No. 1475 (later named *Flying Fox*), was the fourth 4-6-2 to be completed by the LNER, but the first to be officially photographed, in April 1923. It would have been identical with No. 4472 as originally built except for the final digit on the tender. This shows the original paint and lettering style carried by 4-6-2s Nos 1472–81.

**BELOW** No. 4472 when newly repainted at Doncaster in April 1957 and seen as BR No. 60103, in standard British Railways dark green and featuring the then newly introduced second BR emblem on the tender. By comparison with the photograph above, this usefully highlights the several subtle physical changes made to the original Gresley 4-6-2 design between 1922 and 1934, particularly with regard to the chimney, smokebox, outside steam pipe casings, dome, valve gear, location of safety valves, cab, and tender, etc. Although the reversing rod is scarcely visible, another more obvious indication of the only recent conversion to left-hand drive is the simultaneous relocation of the vacuum ejector pipe along the left-hand side of the boiler, below the hand rail.

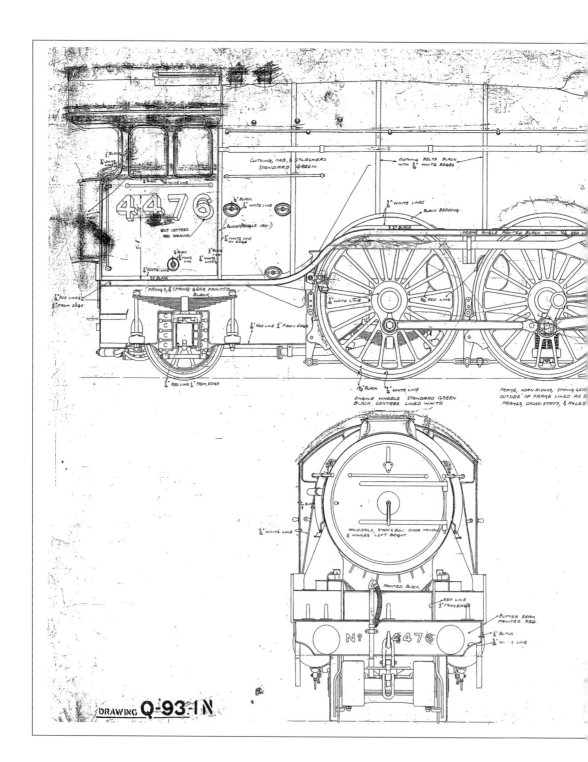

CLOTHING, CAB, & SPLASHERS
STANDARD GREEN

CLOTHING BELTS BLACK
WITH ⅜ WHITE EDGES

¼ WHITE LINES

BLACK BEADING

1¼ BLACK

FRAME ANGLE PAINTED BLACK WITH ½ RED LI

4476

GILT LETTERS
RED SHADING

BLACK

WHITE LINE

BLACK (HINGLE IRON)

WHITE LINE
ON EDGE

WHITE LINE

BLACK
WHITE
LINE

WHITE
LINE

WHITE LINE

⅜ RED LINE

⅜ WHITE LINE
TO BLACK

SPRINGS, & SPRING GEAR PAINTED
BLACK.

RED LINES
FROM EDGE

RED LINE ½ FROM EDGE

RED LINE ½ FROM EDGE

1½ BLACK

WHITE LINE

ENGINE WHEELS   STANDARD GREEN
BLACK CENTRES   LINED WHITE

FRAME, HORN BLOCKS, SPRING GEA
OUTSIDE OF FRAME LINED AS S
FRAMES, CROSS-STAYS, & AXLES

WHITE LINE

HANDRAILS, SMOKE BOX DOOR HANDLE
& HINGES LEFT BRIGHT

PAINTED BLACK

RED LINE
FROM EDGE

BUFFER BEAM
PAINTED RED

Nº 4476

BLACK

WHITE LINE

DRAWING Q·93·1N

HANDRAIL TO BE LEFT BRIGHT.

SMOKEBOX, & CHIMNEY PAINTED BLACK.

CYLNDLR. & CLOTHING PAINTED BLACK

3/16" RED LINE, 1/2" FROM EDGE

ON. BOTTOM EDGE.

GEAR &c. PAINTED BLACK WITH 3½ RED LINE ED RED.

1½" BLACK 3/8" WHITE LINE

## L. & N. E. R.
### DONCASTER

- 6 - 2 THREE CYLINDER EXPRESS PASSENGER ENGINE.

WHEELS 6-8 DIAR   CYLINDERS 20 x 26.   BOILER 6-5 DIAB x 19-0

SCALE 1 INCH = 1 FOOT.

DRAWING N° Q·931N

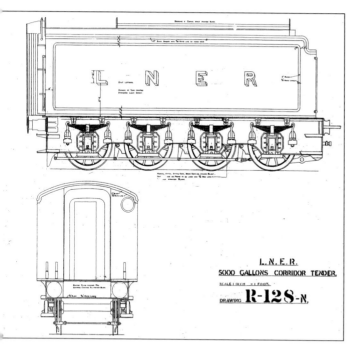

L.N.F.R.

5000 GALLONS CORRIDOR TENDER.

SCALE 1 INCH = 1 FOOT.

DRAWING R-128-N.

**LEFT** Corridor tender paint scheme drawing. From early 1928 both the LMS and LNER ceased to portray a locomotive's running number on its tender, recognising that in practice tenders were often swapped around between engines. Thereafter, tenders simply carried the company insignia.

Just as former Great Northern Railway locomotive practice came to dominate the new London & North Eastern Railway after 1922, GNR passenger locomotive light or 'grass' green was adopted in preference to the Saxony green formerly employed by the 'senior partner', the North Eastern. More recently, this has often been referred to as 'apple green'.

LNER apple green and BR blue: bold lining was black, highlighted on each side with a fine white line. BR dark green: bold lining was black, highlighted with a fine orange line on each side. The shade of green was identical with that employed by the former Great Western Railway, i.e. mid-chrome green, often erroneously described as 'Brunswick green'.

**ABOVE** In 1948 the Railway Executive resolved to paint blue 'the principal express passenger locomotives' of the newly established British Railways. This very fine Gauge 0 scale model depicts No. 4472 in this short-lived guise, as BR No. 60103. It was made by Raymond Walley from a brass kit by David Andrews, and painted by Dennis Morley. On completion in March 2006, it was presented by George Hinchcliffe, on behalf of the Gauge 0 Guild, to the National Railway Museum, where it is displayed in the permanent 'The Flying Scotsman Story' exhibition.

**RIGHT** Tony Filby, National Railway Museum object painter from 1975 until 2008, draws out a template for the tender lettering in September 2006. **FAR RIGHT** Tony Filby applies the finishing touches to No. 4472's driving wheels in March 2007.

| NO. 4472 LIVERY AND NUMBERING CHANGES SUMMARY, 1923–63 | |
| --- | --- |
| **PRE-1963** | |
| February 1923 | Apple green, tender lettered L&NER 1472, engine nameless |
| March 1924 | ditto, tender lettered LNER 4472, engine named *Flying Scotsman* |
| April 1928 | ditto, cabside numbered 4472, tender lettered LNER |
| April 1943 | Black (unlined), tender lettered NE |
| January 1946 | ditto, engine renumbered 502 |
| May 1946 | ditto, engine renumbered 103 |
| January 1947 | Apple green, tender lettered LNER |
| March 1948 | ditto, engine renumbered E103, tender lettered British Railways |
| December 1948 | ditto, engine renumbered 60103 (smokebox door numberplate) |
| December 1949 | Blue, British Railways lion & wheel emblem on tender |
| March 1952 | Dark green, ditto |
| April 1957 | ditto, new British Railways (1956) emblem on tender |
| **POST-1963** | |
| Liveries applied: | |
| March 1963 | LNER apple green, numbered 4472 |
| July 1993 | British Railways dark green, numbered 60103 |
| June 1999 | LNER apple green, numbered 4472 |
| May 2011 | Matt black, lettered NE and numbered 103 and 502 |
| February 2016 | British Railways dark green, numbered 60103 |

## NAMING

Of the 52 A1 4-6-2s as originally turned out, 48 initially entered traffic without names, an omission which had not immediately been addressed. The late Willie Yeadon, chronicler supreme of LNER locomotive minutiae, discovered a written instruction dated March 1925, and issued by Nigel Gresley at King's Cross to William Elwess at Doncaster, that indicated the names which were to be allocated to Nos 4472–81. Gresley remarkably overlooked the fact that No. 4472 had already been given a name just a year earlier, which had particularly captured the public imagination. His new, allocated name for No. 4472 was *Flying Fox*, which in the event was bestowed upon No. 4475.

**LEFT** A close-up of the *Flying Scotsman* nameplate together with its plaque commemorating No. 4472's 422-mile non-stop run in Australia in 1989.

**RIGHT** The original nameplates fitted to No. 4472 in 1924 soon cracked and were replaced by slightly longer plates. A single and unfortunately undated photograph of unknown origin of No. 4472 is reproduced herewith, which clearly shows the engine in post-1928 condition temporarily devoid of nameplates (at least on the left-hand side), leaving York on a down express.

CHAPTER SIX

# The first 40 years, 1923–1963

## From (GNR) No. 1472 to BR No. 60103

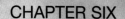

Early days. As No. 1472, and as first turned out without a name, coupled to its original tender inscribed L&NER, the engine having just departed from London King's Cross is seen at Belle Isle with a northbound express sometime in 1923.

ABOVE An extract from the ex-North Eastern Railway dynamometer car log book for June–July 1923, which includes rare official references to No. 1472 as such. (No. 1473 had to be substituted for No. 1472 during 26–27 June.)

## Debut

Having already been cut, the frames for the third Doncaster 4-6-2 locomotive, the would-be Great Northern Railway No. 1472, were laid on 23 October 1922. The engine was considered to be operationally complete 15 weeks later, on Wednesday, 7 February 1923, by the newly constituted London & North Eastern Railway, which had come into being on 1 January. It had cost £7,614 (equivalent to about £340,000 at 2013 price levels). By comparison, the earlier assembly of No. 1470 had occupied 31 weeks. The overall responsibility for both of these complex operations lay with Francis Wintour, Nigel Gresley's deputy and works manager, also late of the Lancashire & Yorkshire Railway.

No. 1472 significantly differed from its two predecessors only in regard to the substitution of laminated for helical springs under the main driving axle. There is no record of its first few days, when it would have run trials painted in plain grey primer. Having performed satisfactorily it would have returned to the paint

shop to be repainted in lined 'apple green', but simplified from the paint scheme applied to GNR Nos 1470/1.

On 22 February, No. 1472 was the largest of 13 locomotives, which collectively represented the recently departed North Eastern, North British, Great Eastern, Great Central and Great Northern railways. These were assembled in London at Marylebone station for the newly appointed LNER directors to select the future locomotive liveries to be adopted by the new organisation. Regrettably, on its first visit to London, the brand-new 4-6-2 is not apparent in either of two surviving official photographs of this occasion.

### Into service

Two days later, on 24 February, No. 1472 officially entered regular traffic, and without a name it joined its two predecessors, No. 1470 *Great Northern*, and No. 1471 *Sir Frederick Banbury,* at Doncaster Carr locomotive shed. There it would likewise initially have been mainly employed between London and Doncaster, although often breaking at Grantham on Yorkshire-bound, particularly Leeds, express workings.

## Comparative trials with ex-NER 4-6-2

When fully run in, in mid-summer 1923, No. 1472 was directly pitted against ex-North Eastern Railway three-cylinder 4-6-2 No. 2400. This had been ordered in March 1922 and hastily completed the following November in response to the GNR 4-6-2s. They participated in a sequence of runs on scheduled workings between Doncaster and King's Cross with the former NER dynamometer car (which is now preserved in the National Railway Museum).

The results were fairly close between the two engines, whose average values are summarised below:

| GNR-TYPE V NER-TYPE 4-6-2 TRIALS, 1923 | | | | | | |
|---|---|---|---|---|---|---|
| Loco No. | Average. dbhp* | Coal lb/dbhp/hr | Coal lb/mile | Average boiler pressure, lb | Average steam temperature, °F | lb water per lb coal |
| 1472 (GNR) | 663 | 3.94 | 48.6 | 164 | 547 | 7.47 |
| 2400 (NER) | 673 | 4.29 | 54.4 | 197 | 574 | 7.70 |

*Drawbar horsepower

From these figures, the NER design appeared to be more effective in maintaining its steam pressure (200lb) and temperature, despite its distinctly inferior boiler, cylinder and valve gear design, although at the cost of slightly heavier coal consumption. Although mechanically speaking the two 4-6-2s differed considerably, there were nevertheless a number of close major dimensional similarities between them. Both 4-6-2s had 6ft 8in diameter coupled wheels and 26in piston stroke, but of the two engines, No. 1472 made a slightly faster showing on the long climb from King's Cross to Potters Bar summit. Despite three more NER-type 4-6-2s being built in 1924, all five engines were scrapped prematurely during 1936–37, when they required new boilers.

## Imperial exhibit

A limited number of photographs exist of No. 1472 as such in service during 1923, towards the end of which it sustained a fractured middle piston rod that could not immediately be replaced. Seemingly, for this prosaic reason alone, it was selected to represent the LNER at the forthcoming British Empire Exhibition, which was due to open at Wembley in north London in May 1924.

No. 1472 entered Doncaster Works for general repairs immediately after Christmas in December 1923. It emerged at the beginning of March 1924 renumbered 4472, and bearing cast brass nameplates inscribed *Flying Scotsman* attached to the middle coupled wheel splashers. The splashers themselves were also embellished with brass beading, which, like the steel wheel tyres were very highly polished, although Gresley insisted on a slightly matt green paint finish for the engine and tender.

The rarely used LNER crest was uniquely applied to the cabsides, while the cylinder covers were chrome plated, and the external brake rigging and external damper rods were bright steel. The engine was despatched south, enveloped in a specially tailored white shroud, and did not return to revenue-earning service until the very end of November 1924. After only four months in traffic it underwent heavy repairs before returning to Wembley for further display during May to October 1925. This time

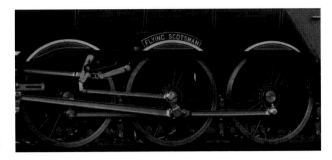

**ABOVE** Detail of No. 4472's special exhibition finish for the 1924 British Empire Exhibition.

**BELOW** No. 4472 as it appeared at the 1925 British Empire Exhibition, with short, six-wheel tender.

**BELOW** No. 4472 as fully refurbished at Doncaster Works in April 1928 preparatory to working the new London–Edinburgh non-stop service.

it was paired with a shorter, six-wheel tender off a K3 class 2-6-0 owing to spatial constraints. Thus, for almost the first three years of its life No. 4472 enjoyed a very cosseted low-mileage existence indeed.

## London–Edinburgh non-stop

No. 4472 finally got down to business, still operating from Doncaster depot, during 1926–27, until it was called into the nearby Plant Works in March 1928. It then lost the cabside crests owing to the new application of cab numerals, but retained the polished brass beading around the coupled wheel splashers, probably until July 1938.

Although only five years old (the best part of 12 months of which had been spent on display at Wembley), it was then fitted with a new (180lb) boiler and new outside cylinders, and also provided with improved, long-travel

valves and valve gear, together with the revolutionary new corridor tender. At the same time, the height of the boiler mountings and cab roof were reduced to conform to the LNER composite loading gauge to permit operation in Scotland, the shorter chimney actually enhancing the engine's overall appearance.

Thus refurbished, No. 4472 was reallocated to King's Cross depot ('Top Shed') in readiness to work the new London–Edinburgh non-stop service due to be inaugurated on Tuesday, 1 May 1928. After seemingly only minimal advance publicity, nevertheless amid great civic pomp on that day, at 10am precisely, it departed north, 'first stop Edinburgh'. It probably crossed the border for the first time, ironically four years after being named! Simultaneously, Glasgow-built No. 2580 *Shotover* (newly rebuilt to A3 with 220lb pressure boiler) steamed south from Edinburgh Waverley with the up train.

The two trains crossed near Pilmoor, not

far north of the mid-way point at Tollerton where ace Gateshead driver Tom Blades would have taken over No. 4472's regulator from Albert Pibworth. As a young fireman on the North Eastern Railway over 30 years earlier, Blades had also participated in the so-called 'Railway Races' of August 1895. No. 4472 and train reached Waverley nine minutes early, at 6.02pm, with more than 2 tons of coal still on the tender, returning a coal consumption of less than 40lb per mile.

## Icon

No. 4472 *Flying Scotsman* was now unquestionably the iconic locomotive of the LNER, Britain's second largest railway company, which at that time, had some 7,400 steam locomotives on its books. These annually powered some 110 million train miles, passenger and freight. Until 1933, the leading express passenger locomotive types of the other three British railway companies were merely large 4-6-0s.

None of these other high-profile locomotives, however, would enjoy a further accolade bestowed on No. 4472, i.e. a starring role in a cinema film which even bore its name. *The Flying Scotsman* was made by British International Pictures in 1929 largely on location on the Hertford Loop, with (acknowledged) considerable liberties being taken with long-established operating and safety procedures in the interests of a good storyline. It starred a young Ray Milland and started off as a silent film, only suddenly to acquire a soundtrack halfway through, thus making British cinematic history.

## 100mph with No. 4472, prelude to the A4

During 20–22 February 1934, No. 4472 was again reunited with the ex-NER dynamometer car, on workings between King's Cross and Edinburgh, which also involved A1 No. 2568 *Sceptre*. Nine months later, on 30 November, following the fitting of a 'special blastpipe' at Doncaster Works to provide a freer exhaust, with the dynamometer car included in the formation of a special light train, No. 4472

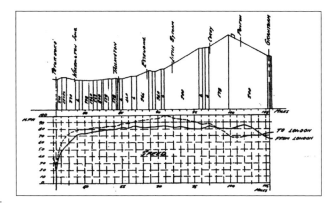

participated in high-speed runs over the 185.8 miles between London King's Cross and Leeds to a schedule of 165 minutes. This was under the command of the legendary driver Bill Sparshatt.

On the initial down journey with a light 147 tons gross, this was completed in merely 151 minutes and 56 seconds, 13 minutes inside the schedule. By this time the engine was almost 12 years old with 653,000 miles under its belt and had covered no less than 44,000 miles since its last general repair only six months previously. Such repairs were undertaken every 60,000 to 80,000 miles at that period, and therefore effectively each year. Nevertheless, despite its exertions earlier in the day, on the corresponding up run with a heavier train aggregating 207½ tons gross, No. 4472 arrived back in London almost seven minutes early and furthermore, touched 100mph at the foot of the 1-in-200 Stoke Bank* near Little Bytham. It thereby became the first British locomotive to achieve an *authenticated* speed of 100mph, albeit only over the course of 600 yards.

This was but the beginning and progressive improvements in Doncaster's successive 4-6-2 designs were reflected in their rising recorded maximum speeds over the next four years:

- 30 November 1934, **100mph***, A1 No. 4472, 8in PV / 20in cyls, 180lb pressure
- 5 March 1935, **108mph***, A3 No. 2750, 8in PV / 19in cyls, 220lb pressure
- 27 September 1935, **112½mph**, at Sandy, A4 No. 2509, 9in PV / 18½in cyls, 250lb pressure, externally streamlined, streamlined steam ports, single exhaust

**ABOVE Speed curves for LNER No. 4472 between Grantham and Peterborough, 30 November 1934.**

**RIGHT** No. 4472 on exhibition at Ilford, Essex on 2 June 1934, together with the brand-new (but by no means spotless) Gresley P2 2-8-2 No. 2001 *Cock o' the North*.

**BELOW** A rare assemblage of close associates (some retired) of Sir Nigel Gresley, several of whom are mentioned in this book, on the occasion of the naming after him of his 100th 4-6-2, A4 No. 4498, at Marylebone station, London, on 27 November 1937. 1. Bert Spencer, 2. Arthur Peppercorn, 3. Francis Wintour, 4. Sir Nigel Gresley, 5. Oliver Bulleid, 6. Harry Broughton, 7. Frank Eggleshaw, 8. Edward Thompson.

■ 3 July 1938, **125mph***, A4 No. 4468, ditto, but with double Kylchap exhaust (125mph was given out at the time, but 126mph was on the plaques fitted to *Mallard* in 1948)

## Decline

An immediate consequence of the A1/A3 high-speed trials was the remarkably rapid design and construction, during mid-1935, of the first four streamlined 'silver' A4 4-6-2s and their matching rolling stock. Initially, these were intended specifically for service on the new high-speed, lightweight 'Silver Jubilee' express

between Newcastle and London. However, by 1938, additional A4s, some equipped with modified second-hand corridor tenders handed down from the A1/A3s, had supplanted these on the most prestigious East Coast services, including the non-stop 'Flying Scotsman'.

No. 4472 lost its corridor tender for the benefit of one of the A4 class in October 1936, and this was initially replaced temporarily by an 'old-style' GNR-type tender, of which photographs showing the engine with this in combination with the 'short' chimney are distinctly rare. This tender was fairly soon replaced by one the most recent non-corridor LNER type in July 1938, which would then remain with No. 4472 for almost 25 years, until the latter's withdrawal from regular service in 1963.

In March 1939, No. 4472 was transferred back to Doncaster from King's Cross, not very long before the outbreak of the Second World War. As a consequence of that conflict, it was repainted plain black in April 1943. Suddenly, during 1944, it experienced a remarkably nomadic life. In rapid succession, it was reallocated to New England (Peterborough) in March, Gorton (Manchester) in July, King's Cross in late October, New England again only two weeks later in early November, and back to Doncaster three weeks later still, in early December.

While remaining black, the engine was renumbered 502 in January 1946, only to be renumbered again as 103 the following May, during a two-day visit to Doncaster Works.

## Post-war years and conversion to A3

No. 4472 was outshopped from Doncaster Works at the beginning of 1947, refreshed in pre-war apple green once more (but with Gill Sans lettering), and also newly converted to Class A3, although initially sporting a conventional round dome. Just over a year later, in March 1948, it received a further repaint, still apple green but lettered 'British Railways' in full on the tender and temporarily renumbered E103, also acquiring a 'banjo' dome boiler for the first time. In December 1948, it received its BR running number of 60103.

In December 1949, it was repainted in the new, short-lived BR blue livery with the new BR 'lion-and-wheel' emblem on the tender, and the following June (1950), No. 4472 was reallocated to Leicester (Central) shed where its duties regularly took it up to London Marylebone and down to Manchester London Road stations.

Wearing blue throughout 1950 and 1951, No. 4472 only returned to Doncaster Works in early February 1952 to emerge the following month repainted in BR standard dark green. At its next shopping during March 1954, it was converted to left-hand drive and so finally mechanically conformed in all respects to the final 1934-batch of A3s. November 1953 found No. 4472 transferred to Grantham for the first time, but it was then briefly moved to King's Cross in June 1954, only to be transferred back to Grantham again two months later. Its final transfer was to King's Cross in April 1957, just after a repaint when the engine had become the first recipient at Doncaster of the newly re-styled BR emblem.

In January 1959, the engine acquired a double chimney, and in December 1961 the distinctive 'German' smoke deflectors, which had been made necessary by the new exhaust system. Almost coincident with the delivery of the first production 'Deltic' locomotives, the recording of the mileage of individual steam locomotives on the Eastern Region officially ceased on 25 March 1961. No. 4472 was then computed to have covered 1,992,757 miles from new. Precisely how accurate this on-going calculation had ever really been is open to question, as indeed it was with any steam locomotive, as these were never fitted with any form of odometer. On withdrawal from BR service nearly two years later, No. 4472 was (unofficially) credited with 2,076,000 miles.

**ABOVE How the mighty have fallen. A rare photograph of No. 4472 in wartime black livery (beneath the grime) in May 1946 newly renumbered 103, and still an A1, working a down express at New Southgate.**

LEFT In the early British Railways regime, with its tender so lettered, and looking considerably refreshed in its second post-war coat of apple green paint, No. 4472 was pictured near Copley Hill, Leeds as BR No. 60103, almost certainly in early 1949. Seen from virtually the same viewpoint as the previous photograph taken in 1946, the intervening conversion to A3 is readily apparent from the square 'blister' on the smokebox side.

**CENTRE** No. 4472 in early British Railways blue livery and numbered 60103, pauses at Grantham on the up 'Flying Scotsman', probably during early 1950 and already be-grimed.

## 1962

By 1962, No. 4472 was fast approaching its 40th anniversary. Diesel-electric traction, which now included the new 3,300hp English Electric 'Deltics', was rapidly taking over the passenger workings on the East Coast Main Line. Nevertheless, in mid-1962, No. 4472 underwent its last scheduled general repair at Doncaster Works, after its emergence from which, the *Railway Observer* reported that No. 4472 remained extremely active during its remaining few months in ordinary service:

### *RAILWAY OBSERVER*, AUGUST 1962

'60103 *Flying Scotsman* (34A) arrived back from Doncaster at the beginning of June. On the first day of the new timetable it was photographed at King's Cross, complete with the "Flying Scotsman" headboard, before working the 10.25am King's Cross–Peterborough and the up "Northumbrian" – still with the headboard! On 2nd July *Flying Scotsman* still in immaculate condition worked a special train to Newcastle for the Royal Show

**LEFT** Ten years later at Peterborough in 1960, BR No. 60103 is seen on the down 'Tees Thames Express', as recently equipped with a double chimney, but not yet provided with smoke deflectors. The blue prototype English Electric *Deltic* is just passing, indicative of the transitional nature of that period.

. . . On the third day of the new timetable, 21st June, the down "Queen of Scots" was headed by 60103 now in its 39th year.'

## RAILWAY OBSERVER, SEPTEMBER 1962

'60103 is still hard at work although, on 27th July, it failed at Three Counties with brake trouble, while working the 1pm ex-Harrogate. The train stood for over an hour before assistance arrived from Hitchin. Four days later *Flying Scotsman* came up nine minutes early with the 5.8pm ex-Hull in place of the "Deltic" which comes off the up "Queen of Scots" at Leeds.'

## RAILWAY OBSERVER, DECEMBER 1962

'King's Cross Pacific, 60103 *Flying Scotsman*, continues to perform in a manner which belies her not inconsiderable age. A week's observations of 60103's workings illustrate the point. On 8th October it worked the 1.15pm King's Cross–Leeds, returning overnight with a fitted freight, the 9th found it on the down "White Rose" and the 4.28pm Doncaster–King's Cross. 60103 worked the 4am King's Cross–Leeds and the up "White Rose" on the 10th, the next day to Grantham with the 5am King's Cross–York parcels. After returning to King's Cross, its next working was the 6.26pm to Doncaster the same day. On the 12th *Flying Scotsman* returned to London with the Edinburgh parcels, her final working for the week was the 4am King's Cross–Leeds and the 12.55pm Leeds–King's Cross on the 13th.'

## Finale

No. 4472's conventional working life finally came to an end during the arctic month of January 1963. At 1.15pm on Saturday, the 14th, with extensive TV and press coverage and with a crowd of admirers present, it departed from King's Cross with an express for Leeds. It actually detached at Doncaster, on what was billed as its final run in public ownership. King's Cross shed closed completely after more than a century the following June, when *scheduled* steam working into that terminus officially ended after 110 years.

LNER practice was to stencil the shed abbreviation on the front buffer beam. After

| NO. 4472 SHED ALLOCATIONS, 1923–63 | | |
|---|---|---|
| Shed | Shed code | Date (from) |
| Doncaster | DON | 24 February 1923 (new engine) |
| London (King's Cross) | KX | 11 April 1928 |
| Doncaster | DON | 6 March 1939 |
| Peterborough (New England) | NWE | 12 March 1944 |
| Manchester (Gorton) | GOR | 7 July 1944 |
| London (King's Cross) | KX | 29 October 1944 |
| Peterborough (New England) | NWE | 11 November 1944 |
| Doncaster | DON | 5 December 1944 |
| Leicester (Central) | 38C | 4 June 1950 |
| Grantham | 34F | 15 November 1953 |
| London (King's Cross) | 34A | 20 June 1954 |
| Grantham | 34F | 29 August 1954 |
| London (King's Cross) | 34A | 7 April 1957 (to withdrawal) |

nationalisation, from September 1949, LMS-style shed codes were indicated by small oval cast iron plates attached to the bottom of the smokebox door.

The distribution of the 79 A1/A3s at their peak in early 1935, i.e. just after the last had entered service in February and prior to the advent of the first A4s in September, was: London (King's Cross) 11, Grantham 14, Doncaster 14, Gateshead 19, Newcastle (Heaton) 5, Edinburgh (Haymarket) 10, Dundee (Tay Bridge) 3, and Carlisle (Canal) 3.

The corresponding distribution of the 78 A3s in 1957, on the eve of dieselisation of the East Coast Main Line was: King's Cross 9, Grantham 12, Doncaster 8, Darlington 2, Gateshead 10, Heaton 13, Haymarket 15, Leeds (Neville Hill) 5, and Carlisle (Canal) 4.

**BELOW** BR No. 60103 *Flying Scotsman* at work, believed to be near Peascliffe in June 1962, a few days after its final routine general overhaul at Doncaster Works.
*(Courtesy, Colour Rail)*

**RIGHT** No. 4472 posed on the RCTS 'East Midlander' at an unknown location, 29 May 1965.

**CENTRE** No. 4472 seen at Clapham Junction, south London, later the same day, with headboard removed.

was detectable as Doncaster practice had been to paint outside cylinder clothing black, but at Darlington the tradition was to adorn this with a green panel. This, and the erroneous red background to the nameplates applied at Doncaster in 1963, lasted until repairs at Derby Works in 1973.

On 10 April 1965, when newly outshopped, No. 4472 worked a special train south over the East Coast Main Line for the benefit of the Darlington staff who had been involved in its recent overhaul. It then proceeded to work the annual 'East Midlander' special, organised by the Railway Correspondence & Travel Society, to run from Nottingham to Swindon and return, on Saturday, 29 May. This was routed via south-west London, where there was a three-hour layover in order to take in a visit to the new Museum of British Transport at Clapham, giving a total journey of 358 miles. Presented as a typical example of the workings upon which No. 4472 was at that period engaged, the full itinerary is given opposite.

## Operating problems and the second tender

Such lengthy excursions with steam locomotives on British Railways were becoming increasingly difficult, mainly due to the problematic logistics of water supplies following the progressive removal by then of water cranes and water troughs. This therefore required the strategic provision of mobile watering facilities en route. As early as September 1964 Alan

**RIGHT** No. 4472 coals at Swindon shed prior to making the return run to Nottingham. The cylinder clothing is painted green, following its recent repairs at Darlington Works.

Pegler enquired as to the possibility of acquiring a second corridor tender, and in 1966 he purchased one in very poor condition, which had been attached to A4 No. 60009 *Union of South Africa*, languishing at Aberdeen.

Although acquired for less than £1,000, a further £5,000 was spent converting it to carry water only, thus increasing its capacity from 5,000 to 6,000 gallons. A combined water capacity of 11,000 gallons was then available behind the locomotive and the corridor facility was necessarily retained. The running number was transferred to the second tender from the cabside, which was adorned once more with the LNER armorial device. Darlington Works having closed completely earlier the same year, this additional tender was attached during the course of light repairs at Doncaster Works in September 1966. A test run with it was made south to Barkston on 3 October.

## Non-stop 'Flying Scotsman' reprise

After considerable planning by Alan Pegler, his master stroke paid off, for on 1 May 1968, No. 4472 with its two tenders in tow, reprised its

non-stop run between London King's Cross and Edinburgh, 40 years to the very day after its inauguration. There were heart-stopping moments at three points when adverse signals reduced speed to a crawl with the danger of an outright stop. However, it was still possible and indeed, deemed necessary to replenish water supplies en route from three remaining water troughs, at Scrooby, Wiske Moor, and Lucker.

These had been retained to replenish the early train-heating boilers of the 'Deltic' diesels. However, because the demands of the latter were much less, the water level in the troughs had been correspondingly reduced. On the day, apparently, the valves on each were covertly held down manually to restore the former 'steam' levels until No. 4472 and train had passed over. Still on schedule as far as Tweedmouth Junction, in the event Edinburgh was reached five minutes outside the 7hr 40min allowed, with 1,000 gallons still in hand, indicating that overall consumption had been just under 14,000 gallons. Three days later, a non-stop up run back to London was also made, which arrived 2½ minutes inside the scheduled 7hr 38min.

Only three months later, on Saturday,

3 August 1968, normal steam locomotive operation on British Railways ceased for ever, and only two small British industrial locomotive manufacturers were still prepared to undertake steam work. Very soon after purchasing No. 4472, Alan Pegler had originally planned to take it to the USA as early as 1965. Three years later, in November 1968, the unprecedented bulk of No. 4472 entered the now-demolished works of the Hunslet Engine Company in Leeds, where it was completely dismantled before undergoing boiler re-tubing and various modifications. These included fitting a prominent chime whistle and the provision of an American locomotive bell, to ensure legal compliance in its role of piloting a mobile British Trade Mission in the United States, which had originally been scheduled for earlier that year.

## Visit to USA and Canada

Having already covered 30,000 miles during the course of railtours since the Hunslet major repair, No. 4472 also later underwent a light overhaul, especially to the motion work, at Doncaster during August 1969. It then worked a chartered special train between King's Cross

and Newcastle on the 31st, by which time it was estimated to have covered 120,000 miles since private purchase six and a half years earlier.

It was then despatched from Liverpool Docks to Boston, Massachusetts, where it was further provided with a pilot, buckeye front coupling and electric headlight in order to comply with other US statutory requirements. During the latter part of 1969, with its seven-coach train, which included two Pullman and one observation car, all still vacuum braked, No. 4472 travelled down from Boston to Houston, Texas. The locomotive presented no problems, but many other difficulties were encountered, manfully overcome by George Hinchcliffe. These particularly concerned electrical supplies to the train when it was stationary. For a variety of reasons, as a commercial enterprise, the venture was not the unqualified financial success that had been anticipated.

No. 4472 and train underwent winter storage in Texas before setting off in 1970 for the Steamtown Railroad Museum at Green Bay, Wisconsin, in order to deliver the two Pullmans which had been promised there. In the event, only one of these arrived. Publicity stops were to be made at Fort Worth, Kansas City, St Louis

**ABOVE** No. 4472, with its two tenders, powers the commemorative special of 1 May 1968, on the East Coast Main Line south of Hatfield.

ABOVE No. 4472 in its obligatory 'regalia' of headlight, chime whistle, pilot ('cowcatcher') and front buckeye coupler, for operation in North America, plus auxiliary tender. Here it is seen at the National Railroad Museum, Green Bay, Wisconsin in 1970. *(A. Wappat)*

and Chicago, the original promotional side of the venture having now been abandoned. The schedule was further extended over the Canadian border to Montreal and on to Toronto, where the ensemble was stored for the winter and for much of 1971.

## Change of ownership and return to the UK

Following the completion of repairs in Buffalo in early 1972, No. 4472 and its remaining entourage set out for San Francisco on the West Coast, a distance of 3,300 miles. By this time, Alan Pegler was experiencing serious financial problems at the hands of several creditors, and it appeared the locomotive might become impounded indefinitely in the USA. After having run an estimated 15,400 miles in North America, No. 4472 was stored at Sharpe Army Base, Sacramento from August 1972 until January 1973. Bill (now Sir William) McAlpine generously came to the financial rescue and purchased the A3 outright.

No. 4472 was then covertly, quickly and

quietly shipped from Oakland in January 1973, and having passed through the Panama Canal, eventually arrived back in Liverpool on 13 February. BR's former rigorous 1968 steam ban having been lifted the previous year, the engine then promptly ran under its own steam to Derby Works for major attention. As No. 4472 had now parted company from its second tender, its number reappeared on the cabside, which therefore lost the LNER crest.

During the summer of 1973, No. 4472 worked between Paignton and Kingswear on the Torbay Steam Railway, before moving to a new base at British Steel, Market Overton near Grantham. Here, its stablemate was none other than GWR No. 4079 *Pendennis Castle*, its one-time adversary at Wembley in 1925. Soon afterwards, both locomotives worked in direct partnership, exchanging trains, in a curious echo of the 1925 LNER/GWR locomotive exchanges almost 50 years earlier. This enabled George Hinchcliffe to make valid comparisons and to calculate that the coal consumption of No. 4472 was some 8 per cent less than that of No. 4079!

In early 1975, for economic reasons, the Market Overton facility was obliged to close and No. 4472 was moved to Steamtown Carnforth, the preserved former locomotive depot alongside the West Coast Main Line in Lancashire. After the fitting of a new smokebox it was despatched from there to participate in the Stockton & Darlington 150 celebrations at Shildon.

During the last three days of November 1977 at York, it reprised its starring role in a major cinema film, *Agatha*, which was set in Harrogate in 1926, for which it was thoughtfully 'cosmetically doctored' to masquerade as fellow class members No. 4474 *Victor Wild*, on its right-hand side, and as No. 4480 *Enterprise*, on its left.

## Second change of boiler

By this time, No. 4472 was in urgent need of very major repairs and after filming was completed it almost immediately proceeded to Messrs Vickers at Barrow-in-Furness.

The spare A4-type boiler purchased by Alan Pegler back in 1965 was now fitted, an operation which did not materially alter the overall appearance of the engine. Although discarded, its latter-day A3 boiler was fortunately not disposed of. Repairs were completed in June 1978, and it was again refreshed during March–June 1980, prior to participating in the 150th anniversary of the Liverpool & Manchester Railway celebrations at Rainhill, Lancashire.

## Visit to Australia

Most of 1985 was occupied by an overhaul at Carnforth, No. 4472 having covered no less than 135,000 miles since leaving Vickers seven years earlier. In February 1988, it moved to Southall for further repairs during which an air brake and electric lighting were fitted in anticipation of shipment to Australia for the forthcoming bicentennial celebrations.

The Pacific was despatched from Tilbury on 12 September and unloaded at Sydney on 16 October. It remained in Australia until November 1989, and while 'Down Under' it ran no less than 28,000 miles, visiting every state except Queensland where the rail gauge was 3ft 6in. Of this, 4,000 miles were accrued in merely 25 days at one point, and at 88mph the pre-war Australian speed record for steam traction was allegedly broken.

Much was made of No. 4472 running non-stop from Parkes to Broken Hill, 422 miles in

**ABOVE** A memento from Australia, an Eveleigh depot (Sydney), New South Wales Government Railways shed plate, affixed to the front plate of 4472's tender.

**BELOW** A rare glimpse of the long-concealed unscientific reality of the standard outside steam pipe arrangement on the Gresley A1s and A3s.

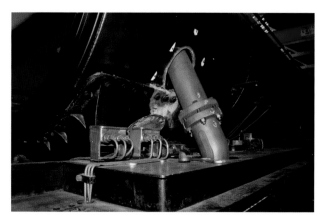

9hr 25min, on 8 August 1989. While this did indeed exceed even the 'emergency' August 1948 408½-mile non-stop runs of the 'Flying Scotsman' in the UK, arguably this still did not constitute a world record. Unfortunately, this overlooked the heroic feat of little Pennsylvania Railroad 4-4-0 No. 573 back in June 1876, i.e. 439 miles in 9hr 57min, as mentioned in Chapter 5.

## Double chimney again

During November/December 1989, No. 4472 was on the high seas again, prior to arriving back at Tilbury. This was immediately followed by a six-month overhaul at Southall in west London, during which the air brake and electric lighting were removed. The engine would be based there for the next 15 years and enjoyed a somewhat nomadic life involving visits to several heritage railways, prior to repairs by Babcock Robey Ltd in Oldbury during April–July 1993.

This included the provision of yet another new smokebox and the somewhat controversial reinstatement of the 1959 double Kylchap exhaust and associated 1961 trough smoke deflectors. This development was made possible by the loan of a double chimney casting formerly carried by none other than No. 4472's nemesis, *Salmon Trout*, which had originally been purchased by a former Southern Region fireman as a garden ornament, for the princely sum of £7 10s (£7.50).

Corridor tender and A4 boiler considerations apart, but otherwise appropriate to the new '1962' configuration, No. 4472 was repainted

in lined BR dark green and renumbered 60103, in response to public demand in a poll taken by *Steam Railway* magazine. In September 1993, Pete Waterman became joint owner of No. 4472 with Sir William McAlpine. After this the 4-6-2 operated on several heritage railways prior to the discovery of major firebox defects in April 1995, while on the Llangollen Railway.

## Under new ownership

Two months later, a major overhaul was authorised, but before this could begin, No. 4472 was purchased by Dr Tony Marchington for £1.2 million in February 1996. A three-year overhaul at Southall, in west London, costing £1 million and supervised by Roland Kennington then ensued. This included replacing the middle cylinder, removing the vacuum brake and re-fitting an air-brake system. This was similar to that provided on the BR Class 37 diesels, with the compressor concealed between the frames ahead of the firebox.

A little-remarked but significant modification, which was also instituted by Kennington, was to install large-bore *direct* steam pipes to the outside cylinders, which remained obscured within the deceptive sheet metal outer casings. These were now 7in bore, which was a very substantial increase over the 4½in of the also more circuitous pipes used in the 1922–30-built 4-6-2s, and which had been increased to merely 5in in the final 1934–35 batch of A3s. The simple analogy of a kink in a garden hose illustrates what an unnecessary impedance to steam flow the original arrangement must have created in so many locomotives for so long.

After this, No. 4472 emerged resplendent once more in iconic LNER apple green and numbered 4472 to satisfy popular commercial demand, despite somewhat incongruously retaining its double chimney and smoke deflectors. Also, in the commercial interest of being able to haul a heavier payload in order to generate more desperately needed revenue to try to recoup mounting costs, an 'invisible' modification was a raising of the boiler pressure to 250lb. Although appropriate to the A4 boiler, strictly speaking this was not valid in an A3, whose three cylinders furthermore had by this time each been bored out to almost the 20in limit. This combination now gave a starting tractive effort of c41,000lb (compared with almost 30,000lb when built in 1923).

While elevating it to the elite '8P' power bracket, this also reduced the adhesion factor to a slippery 3.6, and produced power outputs which still would not have been possible with the original single-chimney arrangement. The result, as demonstrated on the 'Inaugural Scotsman' from King's Cross on 4 July 1999, was spectacular. With a trailing load of 520 tons at 28 per cent cut-off and with the regulator fully open, No. 4472 breasted Stoke Summit at 62mph. Almost nine years later to the day, on 5 July 2008, newly repaired A4 No. 60007 *Sir Nigel Gresley*, hauling 512 tons out of King's Cross, reached Stoke Summit at 69¾mph and at 25 per cent cut-off, from which power outputs of 2,010 *equivalent* drawbar horsepower (edbhp) and 2,490ihp were computed by Douglas Landau. Taking this as a yardstick, in 1999 No. 4472, tempered by its inferior cylinder design, was probably developing something approaching

1,800edbhp and 2,200ihp. Hailed at the time as an A4-style performance, A3½ was perhaps nearer the mark!

By 1999, No. 4472 was, nominally at least, more than 75 years old, and its venerable running gear would not have taken kindly to any regular power outputs of this magnitude. Unfortunately, the huge cost of the most recent overhaul, together with the original purchase price, had set its third and latest private owner back by some £2 million. Four years later, in late 2003, he was sadly declared bankrupt, shortly after No. 4472 had been displayed resplendent at the Doncaster Works 150th anniversary celebration.

The future of No. 4472 was once again in doubt, and amongst several other proposals there were plans to make it the static centrepiece of a new shopping complex at Edinburgh Waverley station. In 2001, Flying Scotsman plc had been formed, but in early 2004 this enterprise offered the locomotive for sale at a price exceeding £2 million, amid fears that it might be lost to an overseas buyer. Tenders were requested by 2 April 2004.

**BELOW No. 4472 resplendent at the Doncaster Works 150th anniversary celebration, July 2003.** *(R.J. Carmen)*

# CHAPTER EIGHT

# 21st-century re-birth

No. 4472 waits at York to depart for Scarborough on the 'Scarborough Spa Express' on 31 May 2005.

# Purchase of No. 4472 for the nation by the National Railway Museum

Following a valuation by Swindon Railway Workshop Ltd, No. 4472 was successfully purchased by the National Railway Museum. This was achieved particularly with the aid of a generous grant of £1.8 million from the Heritage Lottery Fund, but also thanks to very generous contributions from members of the general public amounting to £365,000, a sum which was matched by Sir Richard Branson of Virgin Trains. An appeal organised by *The Railway Magazine* raised a further £50,000.

The purchase of No. 4472 by the National Railway Museum was secured on 5 April, within just two hours of the final deadline. At £2.31 million, the ultimate purchase cost was a far cry from the £3,000 (scrap value) which Alan Pegler had paid for it in 1963. In *real* terms this actually amounted to nearly *five* times its *original* building cost, despite depreciation and the nearly 40-fold decline in the value of the pound sterling over 80 years.

## Arrival at York

In late May 2004 No. 4472 was towed under light steam from Southall to Doncaster, with the intention that it should work a VIP train to York on the opening day of the National Railway Museum Railfest on the 29th. Regrettably, a split fire tube was detected at Doncaster and this 'grand entrance' under its own steam sadly had to be abandoned, although the iconic engine

still received an enthusiastic reception from the assembled crowds.

## Condition of No. 4472 in February 2004

Prior to making an offer, the National Railway Museum commissioned a report as regards No. 4472's mechanical condition from a company representing the Vehicle Acceptance Body (VAB), as the locomotive stood at Southall Railway Centre in West London.

The Vehicle Acceptance Body's report advised that although it had not been possible to complete a thorough examination of the locomotive, or view it in steam, the general mechanical condition of No. 4472 appeared to be satisfactory for continued operation, subject to effective maintenance, until the next scheduled General Overhaul, due in 2006. The report suggested that the scope of mechanical overhaul required was not anticipated to be extensive, but would involve strip down of the motion and axleboxes to gauge wear and any remedial action required, and re-lining and boring of the cylinders to nominal diameter. The report went on to recommend that although the condition of the boiler appeared satisfactory, a thorough boiler overhaul would be required. It was also suggested that any prospective purchaser for the locomotive should secure any available spares as part of the purchase, and should ensure that any spares included were in good condition.

The major spare parts available included both the left- and right-hand outside cylinders ex-*Salmon Trout* (BR No. 60041), whose former middle cylinder had already been installed in

No. 4472 during the 1996–99 heavy overhaul at Southall. These cylinders had presciently been procured by Alan Pegler from the scrappers of No. 60041 in Scotland. There also still remained, almost forgotten and literally put out to grass, the 60-year-old A3 boiler (No. 27020), also previously carried by *Salmon Trout*, and fitted to No. 4472 between 1965 and 1977.

## Under National Railway Museum ownership

With the objective of having No. 4472 available to work the 'Scarborough Spa Express' due to commence on 20 July 2004, the National Railway Museum immediately made the necessary arrangements once the Railfest event was over. All 121 fire tubes (dating from 1999) in the A4-type boiler (No. 27971) then on No. 4472 would be replaced, and the superheater elements refurbished, as well as numerous other remedial tasks. These included replacing three tender springs and certain pipework, servicing the handbrake, and carrying out essential welding repairs on the firebox. The boiler pressure was immediately reduced from 250lb to the historically correct 220lb, to avert the need for even more very expensive repairs, and in order to comply with legal standards. Very quickly, the white metal bearings in two coupled axleboxes and one of the crossheads required particularly urgent attention. A serious crack, which ran the length of the bottom of the right-hand cylinder, and which at some stage had been heavily patched up externally, began to cause concern.

**ABOVE** Coaling No. 4472 at the National Railway Museum. During the steam era, only yards from this location in what is now a museum car park, stood a concrete coaling tower similar to that at King's Cross depot, which is illustrated on page 80, and which was demolished in 1970.

**ABOVE** No. 4472 and train approaches Scarborough on 17 August 2005.

**BELOW** A week later, with its smoke deflector plates experimentally removed, No. 4472 awaits departure from Scarborough on the 'Scarborough Spa Express' on 24 August 2005.

ABOVE No. 4472's chassis, with the nameplates still in place, after the removal of the A4-type boiler at York, February 2006.

ABOVE No. 4472's 'innards' revealed. A close-up of the crank axle and middle connecting rod in situ between the frames.

RIGHT The A4-type boiler (No. 27971) after its removal from No. 4472. It was later sold by the National Railway Museum as a potential spare to Jeremy Hosking, the owner of A4 No. 60019 *Bittern*.

Although the 20 July deadline was met, No. 4472 was not able to work its second scheduled train only two days later, and was failed again a week later due to air-brake problems. Despite these tribulations, No. 4472 managed to put in some good revenue-earning work between York and Scarborough during the summer of 2004, bringing in no less than £100,000 in passenger revenue.

# Final operations with the A4 boiler

With its smoke deflector plates restored and when temporarily based at the Birmingham Railway Museum at Tyseley, No. 4472 operated several Vintage Trains Christmas Lunch specials from Dorridge in the West Midlands in mid-December 2005, with the last on the 17th.

CHAPTER NINE

# The repairs and modifications made to No. 4472 by the National Railway Museum

A high proportion of the original Doncaster working drawings for the Gresley A1/A3 4-6-2s had fortunately survived, and had been given an early priority for cataloguing at the National Railway Museum several years before its purchase of No. 4472. These would greatly assist in the unexpectedly mammoth task which would lie ahead.

## Frames

Remarkably, in view of the traditionally poor frame performance of the A1/A3s in general, and the unnaturally prolonged life of No. 4472 in particular, which had latterly also involved an unprecedented amount of tender-first running, after the frames (of uncertain age) had been stripped down these were found to contain only a single significant crack.

More serious, however, was the discovery that the twin plates were slightly bowed, being $\frac{7}{32}$in (i.e. almost $\frac{1}{4}$in) out of true at the centre of their 42ft length, with the 'hollow' being on the right-hand side. Although seemingly slight and amounting to only one part in one thousand, this deflection could still result in severe wear in both axleboxes and coupled wheel tyres, and cause heated bearings. This defect was corrected by hammering the frames with a riveting tool, although five years later in 2011, a slight joggle was now detected in the left-hand frame between the leading coupled axle and the middle cylinder.

In September 2011, disturbing cracks became apparent in two of the three transverse cast steel stretchers between the frames, which were located between the middle cylinder (which also acted as a substantial stretcher) and the firebox. These afflicted the middle motion plate, which was located roughly above the leading coupled axle and held the slidebars for the middle cylinder, and the particularly large casting situated between the driving and trailing coupled axles. Adequate repairs could not be effected by welding, even after removal of the boiler, and the cost of producing new castings would have been prohibitive.

Following discussions with the VAB, the latter permitted their replacement with fabricated items to be welded up from steel plate, which were then made and installed at Bury. At the same time, it was decided to replace several other lesser stretchers, and generally reinforce the frame structure with steel plates as before.

When removed in late 2012, it was

FAR LEFT An interesting later overview of No. 4472's chassis in the National Railway Museum workshops, taken in 2007.

LEFT The old frame stretcher is lifted out of the frames, 28 November 2011.
*(Chris Birmingham)*

found that each of the cylinders still required significant repairs to major cracks. They were also extensively machined externally, as a result of which the overall width of the middle casting was by now reduced by 0.33in from when originally installed, while the inner faces of the two outside cylinders each 'lost' ⁵⁄₃₂in. Furthermore, each cylinder casting incorporated 50–60 bolt holes which in many cases, did not tally very closely with the corresponding holes drilled in the main frame plates, some of which had become badly elongated due to stresses and strains over the passage of time. The holes in the cylinders were therefore plugged and new ones drilled as per the original working drawings.

Given these anomalies and the fact that First Class Partnerships would not countenance the use of shims to make up the increased fractional differences due to the further machining, the consultants advised that the only viable solution was the outright replacement of the front 12ft portions of both main frame plates.

The steel was ordered from a manufacturer in Finland, which rolled the plates to a thickness of 40mm (1¾in) and supplied them to the main contractor, Arthur Stephenson Engineers Ltd of Atherton, Manchester. The latter then profiled these and uniformly planed them down to the long-standing A3 thickness of 1⅛in, except at the interfaces with the cylinder castings. These were therefore left proud by 0.165in on the inside faces to accommodate the middle cylinder, and by 0.158in on the outside faces in order to preserve the 6ft 8½in lateral spacing between the centres of the two outside cylinders.

ABOVE The fabricated replacement stretcher in course of being welded up, 26 October 2011.

LEFT The new stretcher is about to be lowered into the frames, 16 December 2011. Note the temporary frame spacers.
*(Chris Birmingham)*

**ABOVE** Right-hand view of new front frame section produced at Atherton, near Manchester. The unusual finish is due to the machining operations to achieve the differential thicknesses required to accommodate the middle and outside cylinders. Bury, 3 December 2014.

**LEFT** The new front end frame section highlighting the highly polished increased thickness in the region where the outside cylinder is to be attached. Already mounted is the also newly fabricated smokebox saddle.

**BELOW** Rear view of new front frame section prior to attachment, showing the middle cylinder casting already in place.

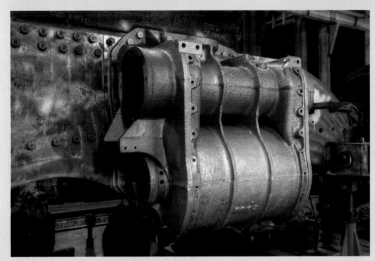

**TOP** The new front end frame section newly welded to the original frames. Bury, 11 December 2014.

**ABOVE LEFT** Grinding down the welded join between the new front frame section and the original frames.

**ABOVE RIGHT** Right-hand outside cylinder newly mounted on new front frame section. Bury, 5 March 2015.

**LEFT** Rear view of right-hand cylinder, showing casting details.

The new frame unit, costing around £80,000, was completed in November 2014, when all three cylinders were temporarily affixed as a check. The old front frame portions were then cut off in early December, immediately after which the new sections were welded on.

## Smokebox saddle

The smokebox saddle was possibly an original casting which dated back to 1923, but after the boiler had been lifted off this was found to be very badly wasted and distorted. Despite the initial repairs made at York, in 2014, an entirely new saddle was fabricated and fitted in conjunction with the new front frame sections.

## Bogie

This was initially considered to be sound, but in 2011 it was decided to provide a new main stretcher, which was cut from solid steel. One new axlebox was also made.

**ABOVE** Rodney Lytton highlights the parlous state of the original smokebox saddle casting, York, July 2006.

**BELOW** The new bogie frame stretcher.

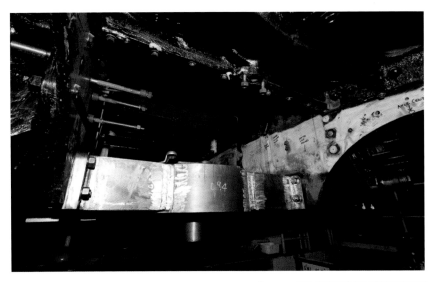

## Hornguides

In June 2011, hairline cracks were discovered in the six coupled axle hornguides, which required their removal and heavy repair. After refurbishment these could no longer be *bolted* back into the frames LNER fashion, as with wear and tear the bolt holes in the frames had increased from 1in to 1¼in diameter. The VAB agreed that these could now be *riveted* in position as per former LMS practice.

LEFT A hornguide is removed from the frames. Note the 20 individually numbered bolts and their corresponding holes in the main frame.

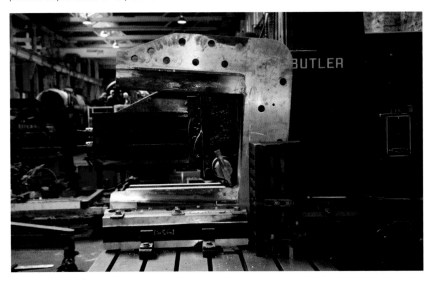

LEFT A newly repaired and fully machined hornguide.

## Cylinders

By the time No. 4472 temporarily ceased operation in December 2005 under its new National Railway Museum ownership, its heavily patched right-hand cylinder was considered to be beyond repair. Fortunately, a spare cylinder ex-*Salmon Trout* was available, which was sent for refurbishment by Corus Northern Engineering Services UK in Workington. (In 2009 Corus was taken over by Tata Steel, which happily continued the former's policy of very generous assistance to the National Railway Museum with respect to No. 4472.)

When mounting the replacement cylinder, it was found that the numerous 1in diameter bolt holes in the cylinder casting did not correspond precisely with the corresponding holes in the frame plate, which had nevertheless matched its predecessor. These slight but significant differences were seemingly satisfactorily reconciled *at the time* by the use of an intermediate steel template, and by inserting dowels into certain holes while slightly enlarging others.

However, in late 2011, it was determined that the middle cylinder was slightly out of alignment in both the vertical and horizontal planes. Two years later, all three cylinders were removed for repairs to cracks by stitch welding, and cast iron liners, chilled with liquid nitrogen, were inserted to reduce their bore to 19in. Their bolt holes were also plugged and re-drilled before they were re-mounted on *new* front main frame sections.

**ABOVE** The left-hand cylinder after repairs at Bury, December 2014.

**RIGHT** The right-hand cylinder being worked on at Corus, Workington, September 2006.

**RIGHT** Re-boring the right-hand cylinder at York on its return from Corus, December 2006.

**BELOW** The front view of the middle cylinder showing the valve gear linkage to its piston valve.

**BELOW The rear view of the middle cylinder.**

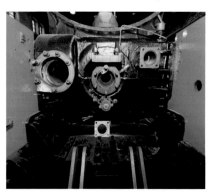

**RIGHT** National Railway Museum object cleaner Barbara Bissell polishes a radius rod from the outside Walschaerts valve gear off No. 4472.

## Valve gear

Although the valve gear itself on No. 4472 required very little attention, new outside motion brackets for each side of the engine were fabricated at Bury in 2011. These were flame-cut from steel plate rather than cast, as had been their predecessors.

## Wheels

In 2006, the driving and coupled wheel centre castings from No. 4472 were found to be very badly pitted, and a specialist spot-welder was engaged to rectify them. Although Doncaster ascribed a life of 60 years to locomotive wheel centres, these bore 'surprising' cast dates of 1948 and 1949. Coincidentally or otherwise, these years corresponded with the construction period of the 49 Peppercorn A1 4-6-2s, of which a single example, No. 60123 *H A Ivatt*, was withdrawn after sustaining collision damage in September 1962, and broken up at Doncaster Works shortly before No. 4472's restoration there. Possibly its much newer, and similar (but not identical) driving and coupled wheel centres were appropriated by No. 4472 in 1963, when it was also fitted with new plain coupled axles.

These wheels were manufactured in Scotland by the Coltness Iron Company, which 25 years earlier, had also cast No. 4472's present trailing wheels, although it was not

**ABOVE** A coupled wheelset from No. 4472 under repair by a spot-welder at York in 2006.

**BELOW LEFT** A close-up view of a coupled wheel spoke on No. 4472 showing a cast date of 1949.

**BELOW** A corresponding view of the trailing carrying wheelset, showing a 'contemporary' cast date of 1923.

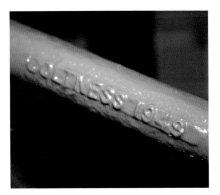

listed as an approved supplier of material in the LNER A1 Specification (see Appendix 7).

In 2010, it was discovered that the back-to-back spacing of the tyres on the trailing coupled wheelset exceeded the specified 4ft 5⅝in by no less than ¼in, whereas the accepted tolerance was a relatively minute 1/64in. The error would appear to have arisen back in 1988 when the coupled wheels had previously been re-tyred. The errant tyre therefore had to be scrapped and a replacement obtained from South Africa at a cost of £6,000, which was fitted at Bury in October 2010 (see page 55). Initially, the new tyre was turned merely to conform with its immediate companion, but a year later all three driving and coupled wheelsets were sent to the South Devon Railway at Buckfastleigh for the six tyres to be turned to a mutually uniform profile.

# Lubrication

Cylinder lubrication was converted back to the original pattern, whereby live steam was taken directly from the boiler for atomisation purposes, rather than indirectly from the steam-operated cylinder drain cocks as hitherto. The lubrication arrangements which served the piston rods, slidebars, and valve spindle glands and guides etc., which involved a considerable amount of fine-bore copper tubing, were heavily renewed.

# Sanding

During the 1996–99 overhaul at Southall, when air brakes were fitted, air sanding had also been fitted to No. 4472 for the first time. The National Railway Museum decided to revert to the more traditional steam sanding in order to reduce the demand on the compressor.

## Suspension

No alterations were made to the suspension arrangements.

## Axleboxes

In 2011, it was found necessary to cast one new bronze coupled axlebox and re-metal the remaining five.

## Brakes

In July 2010, No. 4472's frames were despatched from York to Bury for the necessary modifications to be made and for the new air-brake system to be installed. This differed mainly from that previously fitted in that the locomotive and tender brakes now worked together, whereas before, the tender had worked off the train brake. The braking system was to be linked to the Automatic Warning System (AWS), now supplemented by TPWS, or Train Protection Warning System, that was developed in the late 1990s under Network Rail, which required it to become a mandatory fitting on all locomotives and multiple units by 2004.

A serious disadvantage of AWS had been that a driver could acknowledge a caution signal and yet still proceed past the following stop signal if it was at red, which was not possible with TPWS below a speed of 70mph. TPWS also differed from AWS in that an *audible* warning was no longer required.

For practical and safety considerations, it was considered desirable that the air (locomotive and train) and vacuum (train only where applicable) brake systems should be under a unified control. Such had been devised by Keith Nicholson and it was very similar to that already fitted to *Tornado*, which quite fortuitously, was briefly 'in shops' at York alongside No. 4472 at the crucial time (January 2011). The steam-operated compressor, which was mounted between the frames ahead of the firebox, is of Swedish manufacture, whereas that fitted to *Tornado* is of German origin.

The design of the air-brake system now fitted is to the standard adopted on the main line railways in Britain, known as the two-pipe system. The engine also has a straight air brake for itself

**ABOVE** The new axlebox made at Bury in 2011, mounted on the driving axle.

**ABOVE** Fitting the new air brakes between the frames, Bury, November 2010.

**LEFT** The air brake compressor.

and the tender, used when running light-engine, or for trains not fitted with continuous brakes. The Automatic Air Brake Pipe (AABP) is charged to 72.5lb/sq in and identified by being painted red, the buffer beam hose pipe connections therefore, along with their operating valves, are painted that colour.

The second pipe is the Main Reservoir Pipe (MRP), always fully charged to 100lb/sq in by the compressor, and painted yellow, which keeps the reservoirs on the locomotive and distributed along the train similarly fully charged at all times. Reduction of the AABP pressure by operation of the driver brake valve upsets the distributors on the locomotive and tender, along with all the vehicles in the train, allowing air pressure to enter the individual brake cylinders on the vehicles, applying a force on the train blocks or discs as appropriate. The application is proportional to the level of reduction in the AABP pressure that the driver desires. The reverse takes place when the brake is released.

LEFT The corresponding brake hose connections on the tender drawbar, seen together with the prominent American-style 'Buckeye' coupler.

# Boiler

## The boiler question: to be A3 or A4?

As of early 2004, the respective ages of the spare A3 boiler and current A4 boiler were 59 years and 43 years, although each had seen a similar total of about 30 (not strictly comparable) years of active use. The newer A4 boiler then fitted to No. 4472 retained its original 1960 copper firebox, having seen only barely five years of *regular* (i.e. BR) service before being stored between 1965 and 1977. After this, the A3 boiler (with its *third* copper firebox, which dated from 1957) had been dumped in the open air.

Although the condition of both boilers was undeniably extremely poor, particularly with regard to the state of their respective fireboxes, Jim Rees, then in charge of the National Railway Museum workshops, considered the older A3 unit to be the better of the two, as well as it being historically more authentic. In June 2005, this boiler was despatched to Riley & Son (Electromech) Ltd at Bury in Greater Manchester for heavy repairs.

The single blastpipe and single chimney, reinstated in 1963 and removed again just 30 years later, had apparently since been mislaid. Rees resolved in any case to retain the revived double Kylchap exhaust system in order to ensure reliable steaming with the longer tube A3 boiler upon its return to service, and to allow for any inexperienced firing. With this in view, No. 4472 spent its final week working the 'Scarborough Spa Expresses' from York at the end of August 2005, experimentally running *without* the smoke deflector plates to determine whether these were strictly necessary for compliance with health and safety requirements, i.e. unhindered forward visibility from the footplate. As a result, it was decided that they were indeed desirable.

## A new boiler?

In December 2005, the National Railway Museum briefly examined the possibility of ordering a fully welded all-steel replacement boiler for No. 4472 from Meiningen Works in Germany, which was currently building the dimensionally very closely related boiler for *Tornado*. However, it was

decided that simply replacing the condemned firebox presented a cheaper option.

## The new firebox

The A3 boiler, together with its inner firebox already removed, was despatched to Pridham's Engineering near Tavistock, an enterprise which usefully possessed a 700-ton hydraulic press. In the past, Doncaster and Darlington Works would have cut the very similar A3 and V2 firebox outer casings from a single 9/16in thick sheet of steel initially measuring 22ft by 10ft, and then marked out, punched and drilled all the rivet and pilot stay holes on the flat.

Different techniques were adopted in 2010.

**ABOVE** A3-type boiler No. 27020, fitted to No. 4472 between 1965 and 1977, after sand blasting and repainting in grey primer, is seen prior to despatch to Bury for major repairs to make it fit for further service on the engine, York, June 2005.

**BELOW** The new steel outer firebox casing, at Pridham's.

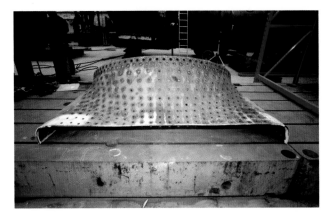

ABOVE The original steel outer throat plate with the bottom 18 inches cut away.

ABOVE RIGHT The new outer steel back plate riveted to the outer casing.

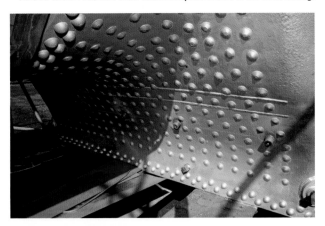

The outer wrapper was now fabricated in three separate subsections, i.e. the pre-formed semi-circular centre portion was welded to the two essentially flat sides, which, as with the copper firebox, were only drilled *after* assembly. The steel back plate was pressed out when red hot, and afterwards riveted to the wrapper; this basic outer firebox alone cost £100,000.

Far more expensive was to be the new *inner* firebox, which presented particular challenges of its own, not least procuring the correct grade of increasingly scarce arsenical copper, which greatly delayed progress. Having once obtained this, fabricating the main structure was relatively simple. What presented particular difficulties, in the absence of the original formers, scrapped in the 1960s, was shaping the three-dimensionally complex inner throat plate.

In 2008, it was estimated that to replace this in one piece would cost £345,000, whilst fabricating it in three parts and welding these together would cost £215,000. A third, and substantially cheaper option (estimated at £66,500), which was adopted, as with the corresponding steel outer throat plate, was to retain the original plate and repair its complex

CENTRE The steel throat plate clearly showing the welded join in place on the new firebox.

LEFT Forming the incandescent copper inner firebox door plate in the 700-ton hydraulic press at Pridham's, September 2007.

upper left- and right-hand shoulders by modern welding techniques. Only the simple, 18in deep 'strip' immediately above the foundation ring was to be replaced.

The original A3/A4 copper fireboxes had been riveted in the traditional manner, but the seams of the new box were welded, a technique for which there existed no up-to-date working standard. On delivery to Bury, defects were unfortunately detected when this was subjected to non-destructive testing (NDT) by Frazer-Nash Consultancy Ltd. This required its return to Devon for rectification by the South Devon Railway, which had in the meantime taken over Messrs Pridham's.

## Assembling the boiler

After attachment of the new outer firebox to the boiler barrel, the boiler was sent from

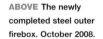

Devon to York, where it was lowered on to No. 4472's frames to check that they were still compatible before work on the boiler progressed further. It was then immediately despatched to Bury, where Riley's had to assemble the firebox, which also required the manufacture of a new foundation ring and a new smokebox tubeplate, before finally fully re-tubing the boiler.

Traditionally, the foundation ring, which united the inner and outer fireboxes, was forged from steel bars, but the new ring was flame-cut in four subsections from 3½in steel plate, which were then welded together.

With the boiler in an inverted position the copper inner firebox, with the foundation ring already attached, was lowered in from above, a once routine but now extremely rare operation, which regrettably, was not photographed at the time. At no point were the copper and steel plates exactly parallel to one another, due to the intervening water space progressively widening from 3½in at the foundation ring to 6in at the boiler 'equator', and the stay holes had to correspond exactly with each other. Following countless cross-referenced measurements, marking these out was therefore a very high-precision operation to tolerances as fine as two- to three-thousandths of an inch, even making allowance for the slight but significant sag in the soft and heavy copper plate.

The pilot stay holes were enlarged and

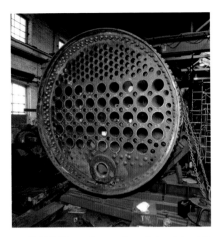

ABOVE The replacement smokebox tubeplate set in the front of the boiler barrel, Bury, January 2011.

ABOVE The life-expired foundation ring from the condemned A3 firebox.

then smoothed, before being tapped at 11 threads per inch, prior to the threaded copper stays being inserted at a strictly specified temperature. (Otherwise the pitch of the threads cut in the plates would be at variance with those cut in the stays as manufactured.) The brackets for the indirect 'sling' or crown stays were salvaged from the old firebox, as were the original longitudinal stays which held the two tubeplates at precisely 19ft apart.

The boiler was re-tubed with tubes and flues to a total value of £50,000, which were kindly donated by Tata Steel. The flues were

ABOVE Hot-riveting the new foundation ring in the inverted firebox, Bury, April 2010.

LEFT The inverted boiler, Bury, October 2010.

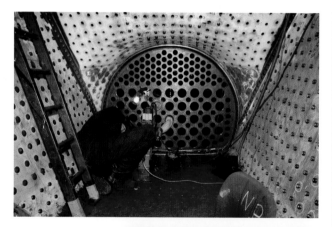

**ABOVE** Work taking place inside the (inverted) firebox. Protective nuts have yet to be applied to the threaded stay ends, Bury, November 2010.

**RIGHT** Checking the flue tubes prior to their installation in the boiler.

**BELOW** The boiler is set up for its hydraulic test at Bury, 17 February 2011.

**ABOVE** The superheater elements, manufactured at a total cost of £14,000, seen en masse, which will later reside within the flue tubes (above left).

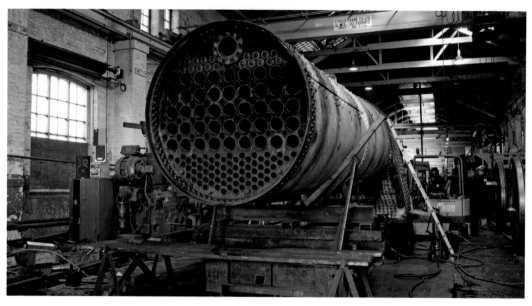

**RIGHT** Checking the integrity of the riveted joint around the firehole, which is the only point where the copper inner and steel outer fireboxes come into *direct* contact with each other.

slightly reduced in diameter at the firebox end, where they were threaded and *screwed* into the copper firebox tubeplate, but *expanded* at their front end in the steel smokebox tubeplate. The smaller-diameter fire tubes were expanded at both ends.

## Testing the boiler

On 7 February 2011, the boiler was again lowered on to the frames at Bury to check that both were still mutually compatible, before being removed to undergo its statutory hydraulic and steam tests.

The hydraulic test, at a pressure of 295lb/sq in with cold water, for a specified duration of 30 minutes, was successfully achieved ten days later. The steam test followed on 21 February, when the boiler was worked first at 230lb, and then 220lb, each for a duration of 30 minutes,

**LEFT** The boiler is signed off after successfully completing its hydraulic test, by boiler inspector John Glaze.

**BELOW** The boiler is fired up for its steam test, Bury, 21 February 2011.

**BELOW** A milestone is reached as the safety valves blow off for the first time (at 230lb pressure). The valves had been newly made to the original specification by South Coast Steam Ltd.

ABOVE The entire Bury workforce poses happily in front of the boiler.

LEFT The new smokebox after its attachment to the boiler, Bury, 22 March 2011.

RIGHT The double chimney casting.

FAR RIGHT Aligning the twin chimneys with the blastpipe centres, Bury, 23 May 2011.

when the newly made Ross 'pop' safety valves lifted for the first time.

A boiler certificate, valid for ten years, could now be issued. With the agreement of the VAB, this would only become activated from the moment a coal fire was first lit on the grate with the boiler mounted on the frames.

## Smokebox

A new smokebox was fitted, the *third* since 1963. The existing cast iron door sufficed for the moment, and there was also a spare. No. 4472 had been provided with a simple spark arrester in the smokebox in 1967. A replacement was now required for the more recent variety, which had fitted round the bulky Kylchap double exhaust system. In late 2015 a new basket-type arrester was devised, closely based on those already fitted to two operating preserved A4 4-6-2s.

## Ashpan

Made from steel plate and subject to high internal temperatures, the existing ashpan required complete replacement. The new unit, costing approximately £10,000, was closely based on the original drawings, but in accordance with Network Rail regulations, a transverse spark

**ABOVE** A view of the new ashpan, Bury, March 2011.

**BELOW** After the ashpan has been laid in the frames the boiler is lowered, Bury, 25 March 2011.

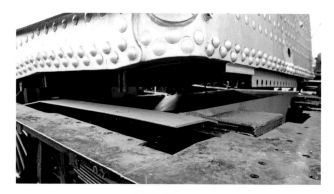

ABOVE **The firebox with its projecting threaded studs is brought down close to the ashpan.**

ABOVE RIGHT **The holes to be made in the ashpan are marked out.**

RIGHT **The holes are cut in the ashpan.**

RIGHT **The nuts are tightened up on the studs to secure the ashpan to the firebox.**

arresting screen has had to be fitted at the front, just inside the damper doors, in order to prevent hot embers falling out on to the track and damaging adjacent signalling cables, etc.

## Superheater

The superheater had yet to be fitted, and the header underwent hydraulic tests at 400lb/sq in pressure at York one week later. The first failed as a flaw was detected in the top of the casting, which was drilled, tapped and then plugged with a short threaded bolt. It underwent a second and successful test a few days later, and was fitted into the boiler at Bury in early May 2011.

BELOW LEFT **The superheater header, with all major exits blanked off, together with the 43 pairs of exit and re-entry points for each element, under hydraulic test at York, 28 February 2011.**

BELOW **Lifting the superheater header into place at Bury, May 2011.**

ABOVE The superheater elements being inserted into the flues and secured into the header.

## Boiler insulation

The original A3 boiler clothing had been disposed of at Barrow in 1978, which required new steel clothing to be made. For the insulation a new, more user-friendly ceramic material was selected in place of the highly unpleasant glass fibre matting, which had latterly been used in succession to lethal asbestos.

TOP Rolling the sheet steel clothing plates, Bury, May 2011.

ABOVE Fitting the clothing plates to the rear of the boiler barrel and firebox.

LEFT The forward part of the boiler barrel was fully clad, flush with the smokebox, prior to the boiler being mounted on the frames.

## Cab

That No. 4472's cab structure, although modified in 1928, was original was confirmed when it was dismantled, revealing hitherto concealed brass strips inscribed '1472' beneath each sliding side window. Unfortunately, the side sheets had become so badly wasted in places over nearly 90 years that they had to be replaced, an operation which was undertaken at York during the winter of 2010–11. The cab area was now required also to accommodate mandatory new state-of-the-art digital equipment, in order to comply with Network Rail regulations:

ABOVE Cutting out the old cab side panels, York, September 2010.

ABOVE RIGHT One of the two hitherto concealed brass strips from inside the cab side sheets inscribed '1472'.

RIGHT Riveting in a replacement side panel in the cab side sheet, Bury, May 2011.

# On-Train Monitoring Recorder (OTMR)

New regulations instituted by Network Rail in January 2006 regarding all locomotives and powered rolling stock, required the installation of an On-Train Monitoring Recorder (OTMR), which has also sometimes more cynically been referred to as 'the spy in the cab'. Manufactured by Arrowvale Electronics of Redditch, at a basic cost per unit of about £6,000, exclusive of the numerous transducers which feed into it, this is akin to the 'black box' fitted in commercial aircraft.

Capable of accommodating up to 120 different channels, this continuously records speeds (which should not exceed 75mph by steam locomotives operating on Network Rail), brake applications and acceleration rates, thereby providing a comprehensive record for retrospective analysis of driver behaviour in the event of any serious transgressions which might have resulted in an accident. On No. 4472 the equipment was fitted beneath the cab wooden floor.

# Global System for Mobile Communications – Railway (GSM-R)

A major development on Network Rail, which has rapidly come into use since 2007, i.e. since No. 4472 was last in active operation in December 2005, and which was scheduled to become operational system-wide by mid-2014, has been the adoption of the advanced digital Global System for Mobile Communications – Railway (abbreviated to GSM-R).

This greatly enhances communications of various kinds between signallers (formerly signalmen) and the drivers of moving trains, even when the latter are adversely located in tunnels or cuttings. The necessary equipment was installed on the front of No. 4472's tender at York in August 2012.

# Whistle

A long-standing method of communication was the humble steam whistle. Despite the highly attractive acoustics of the chime whistle latterly

LEFT The cab takes shape once more. The new air brake control valve (grey) can clearly be seen directly ahead of the driver's seat. Bury, 23 July 2015.

LEFT The GSM-R equipment fitted to the front of the tender, uncovered.

RIGHT The newly installed dial water gauge.

RIGHT The tender front drawbar, showing the massive forged coupler, with brake and water hose connections.

fitted to No. 4472, this was historically incorrect. A standard, former LNER locomotive whistle was therefore purchased at auction and fitted to the engine.

## Tender

Compared with the engine itself, No. 4472's tender required relatively little attention. The tank had been completely renewed in 1982–84, but a new set of eight axleboxes was cast.

For historical interest, in 1984, the old instruction plates for the now-removed water pick-up apparatus had been re-affixed to the front of the new tender tank, but the primitive ty cock method of assessing its water content was retained. This had consisted of a vertical pipe, which was perforated by small holes at intervals. When turned through a right angle, water would spout from these indicating the approximate level of the water in the adjacent tender tank, the front of which carried brass capacity plates set at 1,000 gallon intervals. In 2011, this was supplemented by a simple, non-graduated dial water gauge.

## Painting

### The chimney versus livery question

The two latter-day 1959/61 vintage front end features of double chimney and associated German smoke deflector plates were historically quite incompatible with LNER apple green livery. Nevertheless, in 2011, it was at first decided that in the interests of enduring market appeal, No. 4472 would continue to retain this iconic number in association with apple green livery, notwithstanding the above anomalous features.

### LNER austerity black

As a wartime austerity measure the LNER had repainted the engine in plain *gloss* black at Doncaster Works in April 1943, together with the abbreviated insignia 'N E' on the tender, when it was still numbered 4472. In readiness for its imminent public unveiling at York, in May 2011, No. 4472 was temporarily painted in plain *matt* black at Bury, in conjunction with the short-lived 1946 Thompson numbers of '502'

## Operating No. 4472

**B**oth the reports commissioned by the National Railway Museum concurred that annual mileage should not exceed 6,500 to 7,000 miles, with each individual trip being limited to between 250 and 350 miles. This would result in between 22 and 26 trips per year, 'which would create an appropriate balance between service running for public pleasure and reasonable wear and tear on locomotive moving parts and/or boiler equipment'. First Class Partnerships, employed in an advisory capacity, made recommendations suggesting that future operation on preserved/heritage railways should be limited, due to the likelihood of continuous stopping and starting on twisting lines at gala-type events resulting in an undesirable and disproportionate impact on wear and tear of the locomotive. It was also suggested that in order to avoid inducing undue stresses in the main frames, transport to and from these events by road should only be undertaken in extreme circumstances.

The complicated relationships between the National Railway Museum, the Vehicle Acceptance Body (VAB), the Train Operating Companies (TOCs), and the tour promoters were reviewed. It was recommended that the number of organisations that operate the locomotive should be restricted to an absolute minimum, ideally no more than two, and that

when it was in operation on the national railway network an experienced member of National Railway Museum staff should always be present. Steam raising from cold should be a gradual 24-hour operation in order to minimise stresses in the firebox.

When operating on preserved/heritage railways, this should be restricted to those which are physically connected to the national railway system (to avoid the need for road transport), and be limited to no more than two or three such visits per year. When running at low speeds on such lines the normal lubrication systems, especially regarding cylinders and piston valves, can be less effective. The locomotive should not be driven excessively (including on the national railway network) in order to generate 'pyrotechnic' displays for the benefit of lineside photographers, because of adverse long-term effects on the boiler.

### Maintenance

As regards routine maintenance it was considered that the existing National Railway Museum Locomotive Maintenance Plan dating from 2007, based on the former British Railways system, was adequate. It was even possibly too stringent in some respects, with scope for some adjustment to take account of actual miles run, which was now more appropriate than planning maintenance according to simple time intervals, as had previously been the case.

**ABOVE Taking a well earned rest after that morning's exploits, *Flying Scotsman* is seen on display in the National Railway Museum's North Yard, on 25 February 2016, complete with a Scottish piper.**

Driver Bill Andrew at the regulator of No. 4472 in August 2004.

CHAPTER TEN

# Driving, firing,
# preparation and disposal

## First impressions

There is a particularly evocative account by a prominent Scottish engineman recording his first impressions of Nigel Gresley's early 4-6-2s. Edinburgh's Haymarket shed had received as its initial allocation, the first five Glasgow-built A1s direct from NBL during August and September 1924. The Haymarket engineman and fervent trade unionist, Norman McKillop (who also wrote under the pseudonym of 'Toram Beg'), in his history of ASLEF written 25 years later, *The Lighted Flame* (1950), recalled his unabashed admiration when he first encountered a brand-new Gresley A1 4-6-2:

*'I shall never forget the first time I clapped eyes on one. I simply could not believe it. I had been used to an almost austere looking engine, and here was something that seemed to have all the "works" outside: steam-pipes altering the line from smokebox to footplate, Walschaerts valve gearing cavorting with the greatest abandon all outside the framing, while I had been used to almost every part of the motion being decorously hidden from the common gaze. The huge tapered boiler, the tremendous width of the firebox, leaving little or no room for a footplate, the 5,000 gallon tank on 8 wheels . . . nothing could describe*

*the reaction but that much-abused word, colossal!*

*We went over that engine with a fine-toothed comb, and to us there was only one flaw: Gresley had committed the unpardonable blunder of making his first series of Pacifics right-hand drive. That it was a mistake has been tacitly admitted in the change all his other engines show to left-hand drive. Apart from this it was wonderfully adequate: an almost complete liaison between the Designer and the Engineman.*

*To obtain a proper perspective it is necessary to have a comparison. We had been used for the most part to the Stephenson's valve gear, with inside cylinders. Gresley gave us 99½ per cent of our oiling points comfortably within reach on the outside of the engine. We could hardly believe our eyes when we went underneath and found only a connecting rod and six oiling points to the bogies and driving wheels.*

*We had been used to sitting on a few square inches of wood, either too high or too low to let us see properly. Gresley gave us a padded bucket seat with a back to it\*, and what was more, placed everything within reach. Reverser, brake and throttle handles were just where they should be, and the sensible window was positioned for seeing without our having to corkscrew our bodies or our necks in the process. He gave us a steam chest pressure gauge, which we learned in time to consider almost as reliable as a second watch for working the engine.*

*In short the Gresley Pacifics fulfilled two of the first principles we desired. They were accessible and comfortable to the men who were going to handle them. It was a foregone conclusion that here was an engine that would make its mark.'*

\**Bucket seats were actually provided later, after having first appeared on the V2 and A4 classes.*

This was praise indeed from a Scotsman for a locomotive designed in England. In later years, McKillop became particularly associated with the 1930-built A3 No. 2796 *Spearmint* (later No. 100/BR No. 60100), which spent the greater part of its 35-year working life allocated to Edinburgh Haymarket shed.

# A driver's view (LNER versus LMS 4-6-2s)

**Bill Andrew** retired from the West Coast Railway Company in December 2010 a remarkable 60 years and one month after joining British Railways as an engine cleaner at Rose Grove, near his native Bacup, in November 1950. In the meantime, having served his compulsory National Service (following which many erstwhile railwaymen at that time did not return to work on the railway) he was eventually transferred in 1956 as a fireman to Crewe North depot on the West Coast Main Line.

For the next seven years he regularly fired Stanier former LMS 'Princess' and 'Coronation' four-cylinder 4-6-2s, sometimes as far afield as Perth (320 miles), before becoming one of the last BR firemen to be passed as a fully fledged steam locomotive driver in 1963. By this time, steam locomotives in general, and Stanier 4-6-2s in particular, were rapidly disappearing on British Railways, and during the next few years he would successively drive diesel multiple units, main line diesel-electric and electric locomotives, and even, albeit briefly, the ill-fated Advanced Passenger Train (APT-P).

Against all the odds, 30 years later, Bill would find himself driving not only preserved former LMS 4-6-2s but also former LNER three-cylinder Class A3 and A4 4-6-2s, which placed him in a possibly unique position to judge the relative merits, or otherwise, of the formerly rival East Coast and West Coast Pacific locomotive classes.

Bill first drove No. 4472 along the North Wales coast to Holyhead in 1990 and immediately noted very significant design differences concerning the three principal controls, i.e. the regulator handle, screw reverser and vacuum brake. Regarding the regulator handle, with its alien to-and-fro rather than the more familiar transverse movement, when starting away and opening the regulator valve, in order to avoid slipping, he quickly learned to keep an eye on the steam chest pressure gauge. (This had also been highly appreciated by Norman McKillop more than 60 years earlier, and was a luxury which was not provided on LMS 4-6-2s.)

The reverser in particular also took some getting used to. On the LMS engines the wheel itself worked in the vertical plane, requiring to be wound clockwise for forward gear, whereas on No. 4472 (and the LNER 4-6-2s in general), it was *horizontal* and correspondingly required an *anti*-clockwise turn. Furthermore, whereas on former LMS designs in general, the vacuum brake large ejector was controlled by a wheel, on the LNER its counterpart was a large and heavy suspended lever which would sometimes drop down under its own weight. This could result in an involuntary brake application. Such

**LEFT** Firing No. 4472, September 2005.

an arrangement also required more care to adjust the vacuum in order to achieve a smooth stop, with the vacuum still rising.

Bill also found the gauge glasses to be pitched rather high on the back of the firebox, with the result that it was relatively easy to over-fill the boiler and hence cause priming (the carry-over of water in the boiler and down to the cylinders). He was very surprised that the cylinder drain cock control lever was located on the fireman's side of the cab!

Probably on account of their more massive construction, particularly as regards their frames, Bill Andrew considered the ex-LMS 4-6-2s rode more smoothly than their LNER counterparts, of which the A4 (in his opinion) was superior in this regard to No. 4472. (The A4 had more robust frames, only a single example of which reputedly had ever required outright replacement, while the laminated coupled axle springs were given a slightly longer, 4ft span.)

Interestingly, he considered both LNER designs to be lighter on coal (the A4 in particular, which was also extremely free steaming) than the LMS. They required a lighter fire in the corners of the firebox (grate area 41¼sq ft) compared with the 'Coronation' (50sq ft), which called for a substantial fire to be built up under the door. This was perhaps fortunate as the Doncaster firehole design with its pivoted flap did not readily admit such substantial shovelfuls of coal as on the Crewe-built engines.

Despite its designers' original intentions, Bill found that in practice it was often necessary to drive No. 4472 standing up. In fact, he observed that shorter drivers would find this obligatory in order to work the regulator properly, yet paradoxically, that their smaller stature was a positive advantage when it came to the reverser. Like Norman McKillop, he really appreciated the positive luxury of the padded bucket seat when he *was* able to sit down, in contrast to the spartan 'small wooden squares' grudgingly provided on the LMS 4-6-2s. Although not firing, he missed the American-style steam-operated coal pusher which was (exceptionally) fitted on the Stanier 'Coronations', but appreciated the equally unusual LNER corridor tender for the simple reason that it enabled him

to be provided with cups of tea whilst he remained at the regulator! On the negative side, Bill found No. 4472 to be a rather dirty engine to work on because of coal dust etc. blowing round inside the cab due to a distinct gap which existed between the cab side sheets and the cab floor. This had also been experienced by a regular King's Cross fireman on A3s in general in their final years…

## Firing the A3s, a former King's Cross fireman's recollections

**Dave Rollin** joined Stratford depot in east London as a cleaner in November 1954, and was passed as a fireman only seven months later. In October 1956, he transferred to King's Cross shed and immediately began firing A3s, which for the next 18 months at least, would retain their traditional single exhaust. At first, he had no reason to realise that one or two of them then allocated to King's Cross had recently been fitted with A4 boilers, (Nos 60055 *Woolwinder* and 60101 *Cicero*, soon to be joined by No. 60066 *Merry Hampton*). However, he soon recognised their distinctly improved steaming qualities, even before the general fitting of the double Kylchap exhaust to the A3s from 1958.

It was then found that they could readily be worked at only 15 per cent cut-off, which had previously been something of a rarity, rather than the usual 25 per cent, thus confirming the observation made almost 30 years earlier at the Institution of Locomotive Engineers. He noted, however, that some drivers still persisted in driving by the *sound* of the exhaust, which was now much softer with the Kylchap, and so continued to work the engines at an unnecessarily late cut-off, to the undeserved discomfort of their firemen.

Dave recalled that they would come off shed with a newly made-up fire, which was concentrated at the back of the firebox, at about 150lb boiler pressure when backing on to the train in King's Cross station. On receiving the right away, the fire would be pushed further forward and the damper opened. Fresh coal would be evenly spread around the grate before Gasworks Tunnel, and after emerging from that and before

**LEFT** Dave Burrows throws the traditional burning oily rag into the firebox to 'light up' No. 4472 in 2004.

Copenhagen Tunnel, another 20 shovelfuls would be added, soon followed by a little more.

The exhaust steam injector would be turned on around Holloway, by which time boiler pressure would have reached the full 220lb. Continuous firing at 25 per cent cut-off would be required up to Potters Bar summit, after which things could ease back somewhat. On through runs to Newcastle, at the other end the fire would begin to be run down with no further firing necessary over the remaining 24 miles after passing Durham. He recalled, however, that the hardest turns from King's Cross were to York (188 miles) and Leeds (185 miles) and return, but he considered the general layout of the footplate to be excellent from an operative's point of view.

Dave Rollin was present on No. 4472's footplate on the 1 May 1968 London–Edinburgh non-stop run, and 20 years later he fired the engine during the course of its 422-mile non-stop run in Australia, where he now resides.

## Locomotive preparation

From lighting up, a 4-6-2 boiler containing about 2,000 gallons of cold water took around six to seven hours to reach 50lb steam pressure, at which point the blower could then judiciously be used. However, too rapid a heating-up process (as with cooling down) was undesirable as this could badly stress the firebox via the differential expansion (and

contraction) of the steel and copper plates and so shorten its potential life.

The driver and fireman would book on about an hour before the engine was due off shed. The fireman would first check the water level in the boiler and the state of the fire, progressively building this up as necessary. (On a 4-6-2 15cwt of coal might be consumed before the engine even left the shed.) He also ensured that the fusible plug in the firebox, the tubes and the firebrick arch were all in a satisfactory condition, checked the smokebox and ashpan had previously been cleaned out properly, the smokebox door was tightly shut, and confirmed the dampers operated correctly.

The sandboxes should have been filled, coal stacked safely on the tender, and the fire-

**BELOW** Topping up the forward Wakefield mechanical lubricator with cylinder oil.

irons correctly stowed. The footplate would be kept and swept clear of coal dust etc., with the boiler fittings and the gauge glasses with their armoured-glass protectors in particular all cleaned, after having first isolated the gauge glasses from the pressure in the boiler.

The driver, meanwhile, would also test the water gauges, noting the rising steam pressure and the state of the fire, while also testing both injectors and the brake and sanding gear. Particularly important was his 'oiling-round'

**BELOW** Cleaning the firegrate (with the boiler cold). Note the all-important firebrick arch.

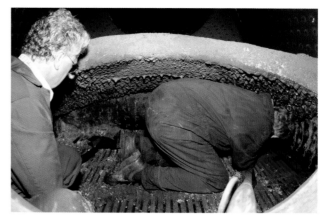

# Appendix 4

## Classified repairs and changes made to No. 4472 at Doncaster Works, 1923-62

| Year | Date | Duration* | Category | Changes |
|---|---|---|---|---|
| 1923 | 7 February (New engine, No. 1472) | - | - | (**B7693** new/**T5223** new) |
| | 16–17 March | 2 days | Light | |
| | 27 December–2 March ('24) | 67 days | General | (1) *(renumbered to 4472, special finish)* |
| 1924 | 10–15 November | 6 days | Light | - |
| 1925 | 23 March–4 April | 13 days | Heavy | **T5378** (2) |
| | 16–28 November | 13 days | Slight | **T5223** |
| 1926 | *no works repairs recorded* - | | | |
| 1927 | 16 February–28 April | 70 days | General | *(variable blastpipe, alteration to buffer beam)* |
| 1928 | 14 February–April | 52 days | General | **B7878** new/**T5323** © (3) *(Knorr piston valves) (soot blower & Colling firebars)* |
| 1929 | 23 April–8 June | 47 days | General | |
| 1930 | 17 January–15 March | 58 days | General | © |
| 1931 | 10 February–2 April | 52 days | General | - |
| 1932 | 6 April–20 May | 55 days | General | - |
| 1933 | 23 February–27 April | 64 days | General | **B7804** © |
| 1934 | 19 April–30 May | 42 days | General | © |
| | 26–27 November | 2 days | Light | *(to fit 'special' blastpipe)* |
| 1935 | 27 March–18 May | 53 days | General | **B7772** |
| 1936 | 25 March–6 June | 84 days | General | © |
| | 19–20 October | 2 days | Light | **T5290** *(to change tender)* |

| Year | Date | Duration* | Category | Changes |
|------|------|-----------|----------|---------|
| 1937 | 25 June–24 July | 30 days | General | |
| 1938 | 13 April | 1 day | | (second-hand bogie fitted) |
| | 27 May–2 July | 57 days | General | **T5640** © |
| 1939 | 18 September–3 November | 47 days | General | **B7785** |
| | 20–23 November | 4 days | Light | (to fit blow-down) |
| 1940 | no works repairs recorded - - - | | | |
| 1941 | 10 May–11 June | 33 days | General | |
| 1942 | no works repairs recorded - - - | | | |
| 1943 | 27 February–3 April | 36 days | General | (black livery) |
| 1944 | 5–23 February | 19 days | Light | |
| 1945 | 3 February–10 March | 36 days | General | |
| 1946 | 18–19 May | 2 days | | (for renumbering to 103) |
| | 18 November–4 January ('47) | 48 days | General | **B8078** (converted to A3, apple green livery) |
| 1947 | no further works repairs - - - | | | |
| 1948 | 2 February–15 March | 43 days | General | **B9119** ('banjo' dome, renumbered E103) |
| 1949 | 4 November–16 December | 43 days | General | **B9448** (blue livery, renumbered 60103) |
| 1950 | no works repairs recorded - - - | | | |
| 1951 | no works repairs recorded - - - | | | |
| 1952 | 5 February–14 March | 38 days | General | **B27015** (BR dark green livery) |
| 1953 | no works repairs recorded - - - | | | |
| 1954 | 8 March–6 April | 30 days | General | **B27074** (altered to left-hand drive) |
| | 13–22 April | 10 days | Unclassified | |
| 1955 | 26 August–8 October | 44 days | General | **B27007** |
| 1956 | no works repairs recorded - - - | | | |
| 1957 | 6 May–13 July | 69 days | General | **B27011** |
| 1958 | 10 December–24 January ('59) | 46 days | General | **B27044** (double Kylchap exhaust) |
| 1959 | no further works repairs - - - | | | |
| 1960 | 8–24 March | 17 days | Casual Light | (AWS?) |
| | 6 July–9 August | 35 days | General | **B27047** (speed indicator?) |
| 1961 | 14 February–4 March | 19 days | Casual Light | - |
| | 21 November–16 December | 26 days | Casual Light | ('trough' smoke deflectors) |
| 1962 | 25 April–2 June | 39 days | General | **B27058** |

**Total days in service 14,587, in works 1,454 days (*including Sundays)** (10.0 per cent).

B = boiler number

T = tender number

© = cylinder change (not recorded after 1938)

**Notes:**

(1) Prepared for display at British Empire Exhibition at Wembley (1924).

(2) Ditto (1925).

(3) Modified for London–Edinburgh non-stop working (long-travel valve gear, corridor tender, and altered to conform to LNER composite loading gauge).

During its 'normal' working life just short of 40 years between February 1923 and January 1963, No. 4472 is recorded as having spent a total of just four years inside Doncaster Works undergoing repairs, and was (unofficially) estimated to have covered a total of 2,076,000 (accident-free) miles.

# Appendix 5

## The official record cards for No. 4472

**LEFT AND BELOW**
LNER record card for No. 4472, 1923–48, obverse and reverse.

**LEFT** LNER/BR record card for
No. 4472, 1945–62, obverse.

**RIGHT** BR record card for No. 60103,
1960–62 obverse.

BRITISH RAILWAYS
THE RAILWAY EXECUTIVE

**ENGINE RECORD CARD**

BR.9215

E. REGN | Region Division | Name | "FLYING SCOTSMAN" | Number 60103

Class (M.R.) 7P | Robinson | Type A3 | Built by Doncaster | Date Feb. 1923

PASSENGER TENDER SUPERHEATER *
MIXED TRAFFIC
FREIGHT TANK NON-SUPERHEATER *

Wheel arrgt. 4-6-2. | Wheel base (E. & T.) 60 ft. 10⅝ ins.

Type of motion Wal/Greeley | Diameter of driving wheels 6 ft. 8 ins.

Engine and Tender—Weight in working order 156 12 Empty 114 12 | Overall height from rail level 13 ft. 1 ins.

No. of cylinders 3 Dia. 19 ins. Stroke 26 ins. Overall length over buffers (E. & T.) 70 ft. 5⅛ ins.

Class of boiler 94 a.

Heating Surface 3398 3434 sq. ft. | No. of tubes { 43 Large Dia. 5¼ ins. / 121 Small Dia. 2¼ ins.

Boiler pressure 220 lbs. per sq. inch. Tractive effort at 85% B.P. 32,909 lbs. Firebox { Steel Copper* } grate area 41.25 sq. ft.

Radius of minimum curve 6 chains (or _____ chains dead slow)   Feed Pump† L & R H. Drive

Brakes† Vacuum Valves† Wilson   Tablet Catching Apparatus†

Injectors (Type & Size) { R.H. _____ / L.H. _____ }   { Cylinders — Atomiser Lubrication / — Sight Feed Lubrication / Axle Boxes — Fountain Lubrication } Mechanical Lubrication†

Axle Boxes (coupled wheels)† _____

C.W.A.* { Front end / Tender end / Both ends } SANDING* { Steam / Mechanical / Back / De-sanding } WATER PICK-UP APPARATUS* { Forward direction / Both directions / Internal fittings only } ROLLER BEARINGS* { Bogie† / Truck† / Coupled† / Tender† } MANGANESE LINERS* { Bogie / Truck / Coupled / Tender }

Drop/Rocking grate _____ Hopper ashpan _____ Self-cleaning smoke box _____ Smoke box deflector plates _____ Spark arrester _____ Blast pipe† Double

Revg. gear { Screw / Lever / Power } Smith-Stone Speed indicator / Speed-Indicator Push and Pull fitted { Vac. C.R. / Air C.R. / Mech. C.R. } Vacuum pump _____ Continuous blowdown _____ Blow-off cock _____ A.T.C.

Trip cocks _____ Condenser _____ Sand gun _____ Gangway doors E & T _____ Storm sheets _____ Back cab _____ Limousine cab _____

Tender weather boards _____ Coal bunker access doors _____ Coal rails _____ Coal pusher _____ Fitted for snow plough _____ Regulator { Smoke box / Dome } Single 3/52

**OTHER NON-SPECIFIED DETAILS:—**

* Delete items not applicable.   † State type.

60103  Back

| | TENDER | | | ALLOCATION | | SHOPS | | | |
|---|---|---|---|---|---|---|---|---|---|
| | | Capacity | | | | | | | |
| Prefix | No. | Water (gall.) | Coal (tons) | Depot | Date | In | Out | Class of Repair | Where |
| E St / BRIT | 5640 | 5000 | 9 | Leicester | 4.6.50 | 13.2.52 | 14.3.52 | GEN. | Doncaster |
| | | | | | | 8.3.54 | 16.4.54 | GEN | Doncaster |
| | | | | | | 26.8.56 | 3.10.56 | GEN | Doncaster |
| | | | | Grantham | 15.11.53 | 6.5.57 | 13.9.57 | Gen | Doncaster |
| | | | | Kings Cross | 20.6.54 | 10.12.58 | 24.1.59 | Gen | Doncaster |
| | | | | Grantham | 29.8.54 | 8.3.60 | 24.3.60 | C(L) | " |
| | | | | Kings Cross | 7.4.59 | 6.7.60 | 9.8.60 | Gen | " |
| | | | | | | 14.2.61 | 4.3.61 | C/L | " |
| | | | | | | 21.11.61 | 16.12.61 | C/L | " |
| | | | | | | 25.4.62 | 2.6.62 | Gen | " |

Experiment Nos. (if fitted)

**ABOVE AND LEFT**
BR record card for
No. 60103, 1950–62
obverse and reverse.

# Appendix 6

## Comparative leading dimensions of Gresley LNER Class A1 and A3 4-6-2s

| Class | A1 | A3 |
|---|---|---|
| Boiler pressure, lb per sq in | 180 | 220 |
| Cylinders | (3) 20 x 26in | (3) 19 x 26in |
| Driving wheel diameter | 6ft 8in | 6ft 8in |
| Heating surfaces, sq ft: | | |
|    Tubes and flues | 2715 | 2422 |
|    Firebox | 215 | 215 |
|    Evaporative | 2930 | 2637 |
|    Superheater | 525 | 706 |
|    Grate area | 41.25 | 41.25 |
| Weights: | | |
|    On leading bogie | 17t 1c | 15t 16c |
|    Bogie | 60t 0c | 66t 3c |
|    Trailing axle | 15t 8c | 14t 6c |
|    Total | 92t 9c | 96t 5c |
| Wheelbase: | | |
|    Coupled | 14ft 6in | 14ft 6in |
|    Total engine | 35ft 9in | 35ft 9in |
| Tractive effort (85% boiler pressure) | 29,835lb | 32,909lb |

**BELOW** Typical LNER Class A1/A3 4-6-2 engine diagram (A3 with original A1 GNR-type tender). No fewer than 31 variants were issued between 1923 and 1948, but none thereafter, e.g. to cover the fitting of A4-type boilers from 1954, or double chimneys from 1958.

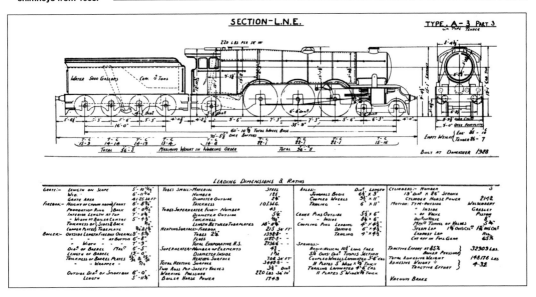

# Appendix 7

## LNER Class A1 4-6-2 Specification (1923)

| APPROVED MAKERS OF MATERIAL | |
|---|---|
| BOILER PLATES & FRAME PLATES | The Leeds Forge Co. Ltd., Leeds |
| | John Spencer & Sons, Ltd., Newcastle upon Tyne |
| | David Colville & Sons, Ltd., Glasgow |
| TYRES & AXLES | Vickers, Ltd., Sheffield |
| | Thomas Firth & Sons, Ltd., Sheffield |
| | Taylor Brothers & Co., Ltd., Leeds |
| | Monk Bridge Iron & Steel Co., Ltd., Leeds |
| CAST STEEL WHEEL CENTRES | The Darlington Forge Co., Ltd., Darlington |
| | Thomas Firth & Sons, Ltd., Sheffield |
| NICKEL CHROME CONNECTING & COUPLING RODS | Vickers, Ltd., Sheffield |
| | Thomas Firth & Sons, Ltd., Sheffield |
| NICKEL CHROME COMBINED PISTONS & PISTON RODS | Vickers, Ltd., Sheffield |
| | Monk Bridge Iron & Steel Co., Ltd., Leeds |
| COPPER PLATES | Williams, Foster & Co., Ltd., Landore |
| | Thomas Bolton & Sons, Ltd., Widnes |
| STEEL BOILER TUBES, FLUE TUBES, & SUPERHEATER ELEMENTS | Howell & Co., Ltd., Sheffield |

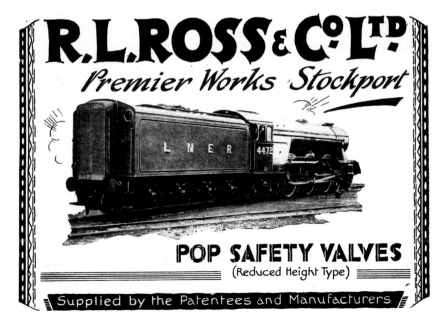

LEFT Trade advertisement, c1929, for R.L. Ross & Co., manufacturers of the safety valves for No. 4472, which was featured in the advertisement.

# Appendix 8

## Commercial models of No. 4472

The author is indebted to Messrs Hornby Hobbies for supplying the three illustrations below and opposite top. Since 1980, Hornby Hobbies Ltd has marketed electric 00 gauge (1/76 scale) models of No. 4472 in numerous different *marques*, which accurately record the various boiler, tender, and livery changes (including BR blue), made during its long career. A live steam version has also been produced. Illustrations of a representative selection of these variants, kindly provided by the manufacturer, are shown below.

**In 1928 condition. Class A1 format with reduced single chimney, round dome casing, and 'plain' smokebox.**
*(Hornby R2675, © Hornby Hobbies Ltd)*

**As restored 1963. Class A3 format, with reinstated single chimney, 'banjo' dome casing, and smokebox 'cover' plates.**
*(Hornby R2261, © Hornby Hobbies Ltd)*

**In 1999–2005 running condition with A4-type boiler (fitted in 1978) with BR double chimney and smoke deflector plates (re-fitted in 1993), and LNER apple green livery (applied in 1999).** *(Hornby R2441, © Hornby Hobbies Ltd)*

**Special limited edition issued in 2011 of the engine as it had then been recently temporarily restored in LNER wartime black livery.** *(Hornby R3100, © Hornby Hobbies Ltd)*

**Bachmann Ltd has produced an electric Gauge 0 (1/43.5 scale) model in brass of No. 4472 in ultimate BR condition with double chimney and smoke deflectors. It is available either unpainted, or painted in lined BR dark green as No. 60103, as here.**

Some years ago, Aster Hobbies Ltd marketed live steam Gauge 0 (1/43.5 scale) models of No. 4472, complete with three cylinders and '2-to-1' derived valve gear. Initially available in 'classic' A3 form in LNER apple green, it later also appeared in the '1962' configuration of BR dark green livery, with double chimney and smoke deflectors.

More recently, Accucraft UK Ltd, which specialises in Gauge 1 (1/32 scale) garden railway equipment, has produced electric and two-cylinder live steam models of the engine in early 1950s, early 1960s, 1975 and 2005 liveries and chimney configurations.

**BELOW Accucraft live steam 1/32-scale model of No. 4472 in classic LNER condition.** *(Accucraft)*

# Select bibliography

## LNER 4-6-2 locomotive development

*Locomotives of the LNER*, Part 2A, Tender Classes A1 to A10, by M.G. Boddy, E. Fry, W.B. Yeadon etc., The Railway Correspondence & Travel Society, 1973

Nock, O.S., *The Gresley Pacifics* Vol.1, 1922–1935, David & Charles, 1973 Vol.2, 1935–1974, David & Charles, 1974

Townend, P.N., *East Coast Pacifics at Work*, Ian Allan Ltd, 1982

## LNER Class A1/A3 4-6-2 locomotives

Reed, B., *Loco Profile 1, LNER Non-streamlined Pacifics*, Profile Publications, 1970

Cook, A.F. Gresley Pacifics and Super-Pacifics, *The Journal of the Stephenson Locomotive Society*, 1981, May, 128–133, June, 159–167, and July/August, 194–198

Yeadon, W.B., *Yeadon's Register of LNER Locomotives*, Vol.1, Gresley's A1, A3 Classes, The Irwell Press, 1990

Coster, P.J., *The Book of the A3 Pacifics*, The Irwell Press, 2003

## No. 4472

Pegler, A.F., et al, *Flying Scotsman*, 3rd Edition, Ian Allan Ltd, 1976

Harris, N., *Flying Scotsman, A Locomotive Legend*, Silver Link Publishing, 1988

Clifford, D., *The World's Most Famous Steam Locomotive: Flying Scotsman*, Finial Publishing, 1997

Nicholson, P., *Flying Scotsman, The World's Most Travelled Steam Locomotive*, Ian Allan Publishing, 1999

Hughes, G., *Flying Scotsman, The People's Engine*, Friends of the National Railway Museum, 2005

Hinchcliffe, G.D., *An Obsession with Steam, the Memoirs of George D. Hinchcliffe*, published privately, 2005

Baldwin, J.S., *Flying Scotsman: The Most Famous Steam Locomotive in the World*, History Press, 2013

## 'Flying Scotsman' train service

The Non-Stop 'Flying Scotsman', *London & North Eastern Railway Magazine*, June 1928, 273/274

**RIGHT** Cover of 31 page booklet entitled *Good Firemanship*, issued by the British Transport Commission (Railway Division) in 1956 as an incentive to reduce locomotive coal consumption on British Railways.

Ellis, C.H., *The Flying Scotsman, 1862–1962, the Portrait of a Train*, George Allen & Unwin, 1962

Mullay, A.J., *Non-stop! London to Scotland in Steam*, Allan Sutton, 1989

Semmens, P.W.B., *Speed on the East Coast Main Line, a Century and a Half of Accelerated Services*, Patrick Stephens, 1990

Gwynne, R., *The Flying Scotsman, the train, the locomotive, the legend*, Shire Publications, 2010

## Nigel Gresley

Nock, O.S., *The Locomotives of Sir Nigel Gresley*, The Railway Gazette, 1945

Brown, F.A.S., *Nigel Gresley, Locomotive Engineer*, Ian Allan Ltd, 1961

Bulleid, H.A.V., *Master Builders of Steam*, Ian Allan Ltd, 1963 (Chapter 2)

Hughes, G., *The Gresley Influence*, Ian Allan Ltd, 1983

Hughes, G., *Sir Nigel Gresley, The Engineer and His Family*, The Oakwood Press, 2001

## Contemporary technical

Specification: British Patent No.15,769/1915, accepted 12 October 1916, *Improvements in valve gear for locomotives and other engines* (derived valve gear for three-cylinder locomotives). H. N. Gresley

Three-Cylinder 4-6-2 Locomotive, Great Northern Railway, *The Railway Engineer*, May 1922, 176–180

Pacific Type Express Passenger Locomotives, London & North Eastern Railway (Great Northern Section), *The Railway Engineer*, March 1923, 95–100 (includes sectional general arrangement drawings)

Specification of a 4-6-2 Pacific type three-cylinder express passenger engine: and eight-wheeled tender, London & North Eastern Railway, King's Cross, November 1923

*LNER Express Passenger Engine 4-6-2 Pacific type 4472 Flying Scotsman: constructed at the Company's works at Doncaster 1922, to the designs of H.N. Gresley*. London & North Eastern Railway, 1924 (reprinted from *Engineering*)

Eagleshaw, F.H., Building the Corridor Tenders, *London & North Eastern Railway Magazine*, July 1928, 342–344

Water supplies for the non-stop Pacifics, *London & North Eastern Railway Magazine*, September 1928, 478–480

Specification: United States Patent No.1,7361,856, accepted 15 October 1929, *Locomotive* (corridor tender). H.N. Gresley

Report: Comparative efficiency of Pacific engines fitted with boilers having working pressures of 180 and 220lb per square inch, and maintenance costs of such boilers. LNER Chief Mechanical Engineer's Department, Doncaster, January 1930

A Record LNER Run, *The Railway Gazette*, 7 December 1934, 951/952 (4472 between King's Cross and Leeds, 30 November 1934)

Cox, E.S., Report on '2-to-1' Valve Gear. LNER 3-Cylinder Locomotives. LMS Chief Mechanical Engineer's Office, Watford, 8 June 1942

Report E26: Dynamometer car trials with Class A3 locomotives on the East Coast Route. British Railways, Eastern Region, 1957

## Restoration of No. 4472, independent reports

Locomotive No. 4472 *Flying Scotsman* Condition Assessment, AEA Technology, February 2004

R. Meanley & R. Kemp, A report for the Trustees of the Science Museum Group into the restoration of A3 Class Pacific *Flying Scotsman* and associated engineering project management, 26 October 2012.

A.D. Roche & A.C. Baker, Final Report of the Independent Review of the Proposed Programme of Works in the Restoration of the *Flying Scotsman* Locomotive 4472, First Class Partnerships, Redacted version, March 2013

## Journals consulted:

*The Engineer, Engineering, Heritage Railway, The Gresley Observer, Journal of the Institution of Locomotive Engineers, Journal of the Stephenson Locomotive Society, London & North Eastern Railway Magazine, Modern Railways, North Eastern Magazine, The Railway Engineer, The Railway Gazette, The Railway Magazine, The Railway Observer, Steam Railway, Steam World, Trains Illustrated, Vintage LNER*

# Glossary

**Adhesion factor** *Adhesive weight* / (starting) *tractive effort*.

**Adhesive weight** The weight which rests upon the coupled wheels of a locomotive.

**Admission** The initial phase of the piston stroke prior to the *cut-off* of the steam supply to the cylinder, after which the steam continues to expand.

**Blastpipe** Orifice from which exhaust steam is discharged inside the smokebox, thereby producing a partial vacuum which creates a draught on the fire.

**Bogie** Four-wheeled truck provided for weight-bearing and guidance purposes.

**Compression** The pressure which results in the cylinder after the valve has closed to *exhaust* before the piston reaches the end of its return stroke.

**Cut-off** The point during the piston stroke at which the *admission* of steam to the cylinder ends (usually quoted as a percentage of the stroke) after which the steam continues to expand.

**Drawbar horsepower** The 'useful' power available at the locomotive drawbar, after a proportion of the *indicated* horsepower has been absorbed in propelling the locomotive itself (abbreviated to dbhp).

**Equivalent dbhp** *Drawbar horsepower* developed on a gradient re-calculated as the equivalent which would have been developed on level track (abbreviated to edbhp).

**Ejector** A simple steam-powered device which extracts air from the locomotive and train vacuum brake system and thereby holds the brakes 'off'.

**Exhaust** The phase of the piston stroke after *expansion* ends and during the return stroke prior to *compression* during which steam is discharged from the cylinder.

**Expansion** The 'working' phase of the piston stroke between *admission* and *exhaust* during which steam continues to expand after *cut-off*.

**Indicated horsepower** The (measurable, i.e. indicated) power developed in the locomotive cylinders, some of which is absorbed in propelling the locomotive (abbreviated to ihp).

**Injector** A simple device which forces feed water into the boiler against its internal pressure by utilising either live or exhaust steam.

**Lap** The amount by which the valve overlaps the steam port when at mid stroke; this permits the continuing *expansion* of the steam in the cylinder after *cut-off*.

**Lead** The amount by which the valve is open to steam at the extreme end of its stroke, thereby providing a cushioning effect on the piston.

**Release** The phase of the piston stroke preceding *exhaust* following *expansion*.

**Superheater** A device which further increases the temperature of the steam as generated in the boiler.

**Tractive effort** More accurately, 'starting tractive effort', or the initial pull exerted by a locomotive moving at negligible speed. For a steam locomotive this was latterly by convention, normally calculated at 85 per cent of the boiler working pressure. (It is not to be confused with horsepower, either indicated or drawbar, and can bear little relation to the *size* of the locomotive.)

**Valve gear** The variable and reversible mechanism which controls the *admission* of steam into and its *exhaust* out of the cylinders under the direction of the driver.

**BELOW** One of the smallest single components on No. 4472, the nevertheless vital BR-manufactured right-hand union link from the outside Walschaerts valve gear.

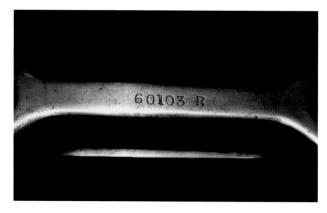

60103 R

# Index